Contents

Publishing Information 1
Bibliographic Key Phrases 1
Publisher's Note 1
TLDR (three words) 2
ELI5 2
Scientific-Style Abstract 3
Semantic Structure 3
    Front Matter 4
    Body 4
    Back Matter 4
Most Important Passages 4
Striking Passages 5

Publishing Information

- (c) 2024 Nimble Books LLC
- ISBN: 978-1-60888-313-4
- version: ADEPT2/v0.21VernorVinge/
- commit 17999454f9406f4cab649b0be577bee654406b77

Bibliographic Key Phrases

uninhabited combat air vehicle; space-based high-energy laser; solar-powered high-energy laser; reconnaissance unmanned air vehicle; attack microbots; piloted single-stage-to-orbit transatmospheric vehicle; uninhabited air-launched transatmospheric vehicle; data fusion; power systems; micromechanical devices; advanced materials; high energy propellants; high-performance computing.

Publisher's Note

In a world grappling with the complexities of rapidly evolving security threats, understanding the future of warfare is paramount. This document, "An Operational Analysis for Air Force 2025: An Application of Value-Focused Thinking to Future Air and Space Capabilities," serves as a vital roadmap, guiding readers through the intricate landscape of air and space power in the coming decades. The document delves into the heart of this challenge, meticulously evaluating a vast array of future system concepts designed to maintain US air and space dominance. It provides a framework, Foundations 2025, for assessing the operational utility of these concepts within six plausible alternate futures, each presenting distinct political, technological, and social challenges.

The analysis uniquely identifies high-leverage technologies crucial for realizing these dominant systems, technologies like data fusion, power systems, and micromechanical devices, many of which have implications extending far beyond the military realm.

"An Operational Analysis for Air Force 2025" is a must-read for anyone invested in understanding the future of warfare, from researchers and practitioners to policymakers and the general public. The document provides a captivating glimpse into cutting-edge military technology, pushing the boundaries of conventional thought and inspiring a deeper understanding of the evolving dynamics of air and space power. The insights offered within its pages will empower readers to engage in informed discussions on crucial defense issues and appreciate the immense potential these future systems hold for shaping the security landscape of the 21st century.

TLDR (three words)

Space LEGO Guide.

ELI5

Imagine you have a big box of LEGOs, but you don't know what to build. You want to build something really cool and strong, something that can fly in the air and in space!

This paper is like a guide to help you choose the best LEGO pieces for your awesome spaceship. It looks at all the different kinds of spaceships people have imagined and tries to figure out which ones are the most important and which LEGO pieces you really need to build them.

It does this by:

- **Thinking about what's important:** Just like you want your spaceship to be strong and fast, the Air Force wants its spaceships to be good at things like finding enemies, reaching faraway places, and fighting. This paper lists all those important things.
- **Giving scores:** Imagine each LEGO piece gets points for how good it is at helping your spaceship do important things. This paper gives points to all the spaceship parts based on how well they do those important things.

- **Imagining different worlds:** What if there are lots of bad guys, or what if spaceships are super easy to build? This paper thinks about different future worlds and how that might change which spaceship parts are the best.
- **Finding the best LEGO pieces:** After looking at all the spaceship parts and how they score in different future worlds, this paper tells you which LEGO pieces are the most important to collect. These are things like special engines, super-strong materials, and tiny computers.

So, this paper helps the Air Force decide which spaceship parts to focus on building so they can be ready for anything in the future!

Scientific-Style Abstract

This research paper reports on an operational analysis conducted to evaluate the operational utility of future air and space system concepts proposed in the Air Force 2025 study. A bottoms-up value model, Foundations 2025, was developed that reflects the values of the Air Force 2025 study participants. Foundations 2025 is unique because it prioritizes a large number of systems and is cast 30 years into the future. Analysis using Foundations 2025 suggests that awareness and space control will be critical contributors to air and space dominance in the future. Furthermore, high leverage technologies identified by the analysis include data fusion, power systems, micromechanical devices, advanced materials, high-energy propellants, and high-performance computing.

This study evaluated 43 future air and space system concepts to identify those offering the greatest potential to enhance US air and space dominance in the year 2025 and beyond. A value model, Foundations 2025, was developed for this assessment. The study also analyzed the technologies required to enable these systems, explicitly taking into account both the importance of the technologies to those systems and the value of the systems to air and space operations. The top five systems were Global Information Management System, Sanctuary Base, Global Surveillance, Reconnaissance, and Targeting System, Global Area Strike System, and Uninhabited Combat Air Vehicle. Six high leverage technologies were identified: Data Fusion, Power Systems, Micromechanical Devices, Advanced Materials, High Energy Propellants, and High Performance Computing. The study concludes that future air and space forces will derive significant advantage from improved awareness and control of space.

Semantic Structure

Parts > {Front Matter, Body, Back Matter} => {Foreword, Chapters, Indexes} => {Scenes, Content}

Front Matter

- **Disclaimer:** Legal/ethical boundaries
- **Contents:** Report organization
- **Acknowledgments:** Thanks to contributors
- **Executive Summary:** Summarizes report

Body

- **Chapter 1:** Introduces AF 2025 study
- **Chapter 2:** Explains analysis methodology
- **Chapter 3:** Presents analysis results
- **Chapter 4:** Summarizes conclusions

Back Matter

- **Appendix A:** Detailed value model
- **Appendix B:** Detailed system descriptions
- **Appendix C:** Alternate futures weights
- **Appendix D:** Technology model
- **Appendix E:** System scores
- **Appendix F:** Technology scores
- **Bibliography:** Source documents

Most Important Passages

> "The Air Force 2025 study produced a number of excellent system concepts for employing air and space power in the future. Analysis of the highest value system concepts indicated that the effort to occupy the "high ground" of the future will require air and space forces to possess increased awareness and to control the medium of space." (xii)

Rationale: This passage states the two most important findings of the report: that improved awareness and control of space will be essential in the future.

> "Analysis of the highest value system concepts indicated that the effort to occupy the "high ground" of the future will require air and space forces to possess increased awareness and to control the medium of space. The five highest value system concepts were: - Global information management system - Sanctuary base - Global surveillance, reconnaissance, and targeting system - Global area strike system - Uninhabited combat air vehicle" (xii)

Rationale: This passage summarizes the findings regarding the systems that offer the most value. Note that all of them emphasize the importance of awareness, and four of the five operate in space.

"The study also included an assessment of the enabling technologies on which the system concepts depend. The analysis explicitly took into account the number of system concepts each technology supported, the degree to which each system concept depended on it, and the importance of the system concept. Six high-leverage technologies stood out because they are important to a large number of high-value system concepts: - Data fusion - Power systems - Micromechanical devices - Advanced materials - High energy propellants - High-performance computing" (xiii)

Rationale: This passage summarizes the findings regarding the technologies that show the most promise.

"The major surprise among these results was the importance of continued breakthroughs in the area of power systems. Other moderate-leverage technologies were also important but contributed to only three or four of the high-value system concepts: - High-energy laser systems - Artificial intelligence - Optics - Aerospace structures - Image processing - Communications" (xiii)

Rationale: This passage continues the discussion of promising technologies and emphasizes that Power Systems scored higher than expected.

"This analysis contends that the high ground of improved awareness offers significant potential for achieving future air and space dominance. Typically, top-scoring systems possessed higher degrees of awareness and/or were predominantly space systems" (66)

Rationale: This passage reiterates the importance of awareness and space dominance and reinforces the relationship between high-scoring systems and those that emphasize awareness and/or are space-based.

Striking Passages

"The long-range planning process in our Air Force is broken. If we are going to be relevant in the future, we've got to somehow break free of the evolutionary nature of the planning process." (1)

"Instead of highly detailed system data, VFT employs user-defined scoring functions to compare the performance of alternate systems. Systems that perform well on these scoring functions receive high-value scores." (9)

"A number of senior decision makers have viewed the model and commented that the best use of Foundations 2025 may be an analysis of systems within the distinct spheres of awareness, reach, and power. They envision separating and developing each function of the model further (refining the tasks, subtasks, force qualities, measures of merit, and scoring functions) and studying which awareness (or reach or power) systems are most promising. These three separate models could be effective mission area analysis tools for the major commands." (70) # Striking Passages

"The long-range planning process in our Air Force is broken. If we are going to be relevant in the future, weâ€™ve got to somehow break free of the evolutionary nature of the planning process." (1)

"Instead of highly detailed system data, VFT employs user-defined scoring functions to compare the performance of alternate systems. Systems that perform well on these scoring functions receive high-value scores." (9)

"A number of senior decision makers have viewed the model and commented that the best use of Foundations 2025 may be an analysis of systems within the distinct spheres of awareness, reach, and power. They envision separating and developing each function of the model further (refining the tasks, subtasks, force qualities, measures of merit, and scoring functions) and studying which awareness (or reach or power) systems are most promising. These ... separate models could be effective mission area analysis tools for the major commands." (70)

An Operational Analysis for *Air Force 2025*: An Application of Value-Focused Thinking to Future Air and Space Capabilities

A Research Paper
Presented To

Air Force *2025*

by

Lt Col Jack A. Jackson, Jr., PhD, AFIT
Lt Col Brian L. Jones, PhD, AFIT
Maj Lee J. Lehmkuhl, DrSci, AFIT

with major contributions from

Col Gregory S. Parnell, USAF Ret., PhD, VCU
Col Gerald A. Hasen, PhD, AFIT
Lt Col Gregg H. Gunsch, PhD, AFIT
Lt Col John M. Andrew, PhD, AWC
Maj Thomas A. Buter, PhD, AFIT
Maj Harry W. Conley, ACSC
Maj Robert F. Mills, PhD, AFIT

May 1996

Form SF298 Citation Data

Report Date ("DD MON YYYY") 00051996	Report Type N/A	Dates Covered (from... to) ("DD MON YYYY")

Title and Subtitle An Operational Analysis for Air Force 2025: An Application of Value-Focused Thinking to Future Air and Space Capabilities	**Contract or Grant Number**
	Program Element Number
Authors Jackson, Jr., Jack A.; Jones, Brian L.; Lehmkuhl, Lee J.	**Project Number**
	Task Number
	Work Unit Number
Performing Organization Name(s) and Address(es) Air Command and Staff College Maxwell AFB, Al 36112	**Performing Organization Number(s)**
Sponsoring/Monitoring Agency Name(s) and Address(es)	**Monitoring Agency Acronym**
	Monitoring Agency Report Number(s)
Distribution/Availability Statement Approved for public release, distribution unlimited	
Supplementary Notes	
Abstract	
Subject Terms	
Document Classification unclassified	**Classification of SF298** unclassified
Classification of Abstract unclassified	**Limitation of Abstract** unlimited
Number of Pages 190	

Disclaimer

2025 is a study designed to comply with a directive from the chief of staff of the Air Force to examine the concepts, capabilities, and technologies the United States will require to remain the dominant air and space force in the future. Presented on 17 June 1996, this report was produced in the Department of Defense school environment of academic freedom and in the interest of advancing concepts related to national defense. The views expressed in this report are those of the authors and do not reflect the official policy or position of the United States Air Force, Department of Defense, or the United States government.

This report contains fictional representations of future situations/scenarios. Any similarities to real people or events, other than those specifically cited, are unintentional and are for purposes of illustration only.

Mention of various programs or technologies throughout this paper does not imply Air Force or DOD endorsement of either the mission, the program, or adoption of the technology.

This publication has been reviewed by security and policy review authorities, is unclassified, and is cleared for public release.

Contents

Chapter		Page
	Disclaimer	ii
	Illustrations	vi
	Tables	x
	Acknowledgments	xi
	Executive Summary	xii
1	Introduction	1
	Meeting the Challenge	1
	Overview of Report	4
2	Methodology	6
	Challenges	6
	The *Air Force 2025* Operational Analysis	7
	Selecting a Decision Analysis Tool	8
	Comparing Analysis Tools	8
	The Best Approach for *Air Force 2025*	10
	Value-Focused Thinking	10
	Value	11
	Objectives, Functions, Tasks, and Subtasks	11
	Force Qualities	11
	Measures of Merit and Scoring Functions	11
	Value Model	13
	Weights	14
	Applying the Value Model	15
	The Search for the *Air Force 2025* Value Model	16
	Developing the *Air Force 2025* Value Model—*Foundations 2025*	21
	Objective	21
	A Bottom-Up Approach	21
	Tasks and Subtasks	21
	The Affinity Exercise	22
	Distinguishing Subtasks from Tasks	25
	The Influence of the Medium on Certain Tasks	25
	Functions	27
	Force Qualities	29
	Measures of Merit and Scoring Functions	32
	System Identification	33
	Alternate Futures	36
	Scoring the Systems	37
	Weighting the *Foundations 2025* Value Model across Alternate Futures	38
	Technology Identification and Scoring	38

Chapter	Page
3 Results	42
Scoring the Technologies	53
4 Conclusions	66
Major Implications of the *Air Force 2025* OA	66
OA Process Lessons Learned	69
Study Limitations	69
Major Implications for the Future	70
Summary	70

Appendix	Page
A *Foundations 2025* Value Model	71
Value Model	71
Measures of Merit and Scoring Functions	75
Operational Analysis Glossary	85
B System Descriptions	89
1.1 Hypersonic Attack Aircraft	91
1.2 Fotofighter	92
1.3 Container Aircraft	93
1.4 Lighter-than-Air Airlifter	94
1.5 Supersonic Airlifter	95
1.6 Stealth Airlifter	96
1.7 Global Transport Aircraft	97
2.1 Strike UAV	98
2.2 Reconnaissance UAV	99
2.3 Uninhabited Combat Air Vehicle	100
2.4 Precision Delivery System	101
2.5 UAV Mothership	102
2.6 Exfiltration Rocket	103
3.1 Orbital Maneuvering Vehicle	104
3.2 Orbital Combat Vehicle	105
3.3 Satellite Bodyguards	106
4.1 Piloted SSTO Transatmospheric Vehicle	107
4.2 Uninhabited Air-Launched Transatmospheric Vehicle	108
5.1 Adjustable Yield Munition	109
5.2 Advanced Air-to-Air Missile	110
5.3 Airborne High-Power Microwave Weapon	111
5.4 Standoff Hypersonic Missile	112
5.5 Attack Microbots	113
5.6 Airborne Holographic Projector	114
5.7 Hybrid High-Energy Laser System	115
6.1 Global Area Strike System	116
6.2 Space-Based Kinetic Energy Weapon	117
6.3 Space-Based High-Power Microwave Weapon	118
6.4 Space-Based High-Energy Laser System	119
6.5 Solar-Powered High-Energy Laser System	120
6.6 Solar Energy Optical Weapon	121
6.7 Asteroid Mitigation System	122
7.1 Spoken Language Translator	123
7.2 Personal Digital Assistant	124
7.3 Virtual Interaction Center	125
8.1 Global Information Management System	126

Appendix *Page*

        8.2 Global Surveillance, Reconnaissance, and Targeting System 127
        8.3 Sensor Microbots 128
        8.4 Multiband Laser Sensor System 129
        8.5 Asteroid Detection System 130
        9.1 Mobile Asset Repair Station 131
        9.2 Weather Analysis and Modification System 132
        9.3 Sanctuary Base 133

C   Alternate Futures Weights 134

D   Technology Model 142
        Level 1 Technologies 145
        Level 2 Technologies 147
        Level 3 Technologies 155

E   System Scores 158

F   Technology Scores 166

    Bibliography 176

Illustrations

Figure		Page
1-1.	*Air Force 2025* Study Process	2
1-2.	Operational Analysis Methodology	4
2-1.	*Air Force 2025* OA Process	8
2-2.	Sample Scoring Functions	12
2-3.	Notional Value Model	14
2-4.	Weights Applied to Notional Value Model	15
2-5.	Value Model Based on JROC/JWCA Categories	17
2-6.	Value Model Based on AFDD-1	18
2-7.	Value Model Based on *Joint Vision 2010*	19
2-8.	Value Model Based on *Cornerstones*	20
2-9.	*Air Force 2025* Initial Task Compilation	23
2-10.	First Affinity Grouping of Tasks	24
2-11.	Second Affinity Grouping of Tasks	25
2-12.	Inclusive Grouping of Tasks and Subtasks	27
2-13.	*Foundations 2025* Value Model	29
2-14.	*Awareness* Tasks, Subtasks, and Force Qualities	30
2-15.	*Reach* Tasks, Subtasks, and Force Qualities	31
2-16.	*Power* Tasks, Subtasks, and Force Qualities	32
2-17.	System Functional Hierarchy	34
2-18.	Alternate Futures Planning Space	36
3-1.	Representative Systems	43
3-2.	Final System Scores - AU Team Weights	45

Figure		Page
3-3.	Final System Scores - Alt. Futures Weights	46
3-4.	Top 11 Systems - AU Team Weights	47
3-5.	Top 14 Systems - Alt. Futures Weights	48
3-6.	Top 11 System Rankings	49
3-7.	*Awareness* Scores - AU Team Weights	50
3-8.	Deploy Scores - AU Team Weights, Halfs Future	51
3-9.	*Power* Scores - AU Team Weights	52
3-10.	Technology Rankings (All 43 Sys, AU Students Wts)	58
3-11.	Top 12 Technology Rankings (All 43 Sys, AU Students Wts)	59
3-12.	Technology Rankings (Top 11 Sys, AU Students Wts)	60
3-13.	Top 12 Technology Rankings (Top 11 Sys, AU Students Wts)	61
3-14.	Technology Rankings (All 43 Sys, Alt Fut Team Wts)	62
3-15.	Top 12 Technology Rankings (All 43 Sys, Alt Fut Team Wts)	63
3-16.	Technology Rankings (Top 11 Sys, Alt Fut Team Wts)	64
3-17.	Top 12 Technology Rankings (Top 11 Sys, Alt Fut Team Wts)	65
A-1.	Value Model - Top Level	71
A-2.	Value Model - Awareness	72
A-3.	Value Model - Reach	73
A-4.	Value Model - Power	74
A-5.	*Detect in Air* Scoring Functions	75
A-6.	*Detect in Space* Scoring Functions	75
A-7.	*Detect in Cyberspace* Scoring Functions	76
A-8.	*Detect on Surface/Subsurface* Scoring Functions	76
A-9.	*Identify* Scoring Functions	76
A-10.	*Integrate* Scoring Functions	77
A-11.	*Educate/Train* Scoring Functions	77
A-12.	*Assess* Scoring Functions	77

Figure		Page
A-13.	*Plan* Scoring Functions	77
A-14.	*Decide* Scoring Functions	78
A-15.	*Communicate* Scoring Functions	78
A-16.	*Confirm* Scoring Functions	78
A-17.	*Deploy to Air* Scoring Functions	79
A-18.	*Deploy to Space* Scoring Functions	79
A-19.	*Deploy to Surface* Scoring Functions	80
A-20.	*Readiness* Scoring Functions	80
A-21.	*Sustain* Scoring Functions	80
A-22.	*Replenish in Air* Scoring Functions	81
A-23.	*Replenish in Space* Scoring Functions	81
A-24.	*Replenish on Surface* Scoring Functions	82
A-25.	*Engage in Air* Scoring Functions	82
A-26.	*Engage in Space* Scoring Functions	82
A-27.	*Engage in Cyberspace* Scoring Functions	83
A-28.	*Engage on Surface/Subsurface* Scoring Functions	83
A-29.	*Survive in Air* Scoring Functions	84
A-30.	*Survive in Space* Scoring Functions	84
A-31.	*Survive in Cyberspace* Scoring Functions	84
A-32.	*Survive on Surface* Scoring Functions	84
B-1.	System Hierarchy	89
C-1.	AU Team Weights - Halfs Future	134
C-2.	AU Team Weights - Gulliver's Future	135
C-3.	AU Team Weights - Zaibatsu Future	135
C-4.	AU Team Weights - Digital Future	136
C-5.	AU Team Weights - Khan Future	136
C-6.	AU Team Weights - 2015 Future	137

Figure		Page
C-7.	Alt Futures Weights - Halfs Future	137
C-8.	Alt Futures Weights - Gulliver's Future	138
C-9.	Alt Futures Weights - Zaibatsu Future	138
C-10.	Alt Futures Weights - Digital Future	139
C-11.	Alt Futures Weights - Khan Future	139
C-12.	Alt Futures Weights - 2015 Future	140
C-13.	*Awareness* Force Quality Weights	140
C-14.	*Reach* Force Quality Weights	141
C-15.	*Power* Force Quality Weights	141
D-1.	Technology Model - Part I	143
D-2.	Technology Model - Part II	144
E-1.	Value Model - Top Level	158
E-2.	System Scoring - *Detect* Task	159
E-3.	System Scoring - *Understand/Direct* Tasks	160
E-4.	System Scoring - *Deploy/Maintain/Replenish* Tasks	161
E-5.	System Scoring - *Survive* Task	162
E-6.	System Scoring - *Engage* Task	163
E-7.	Final System Values By Future - AU Team Weights	164
E-8.	Final System Values By Future - Alt Futures Weights	165

Tables

Table		Page
1	Identified Systems	35
2	Technology Assessment	54
3	Technology Development Leaders	56
4	Writing Team Letter Designators and Names	90
5	Technology Scores by System	167
6	Weighted Technology Scores (All 43 Sys, AU Students Wts)	172
7	Weighted Technology Scores (Top 11 Sys, AU Students Wts)	173
8	Weighted Technology Scores (All 43 Sys, Alt Fut Team Wts)	174
9	Weighted Technology Scores (Top 11 Sys, Alt Fut Team Wts)	175

Acknowledgments

We would like to thank Col George K. Haritos, the faculty of the Graduate School of Engineering, and the faculty of the Graduate School of Logistics and Acquisition at the Air Force Institute of Technology for their contribute to the *Air Force 2025* study. Colonel Haritos formed a team of senior faculty who examined issues surrounding which sectors (DOD or commercial or both) might take the lead in future technology development.

We also would like to recognize the contribution of Lt Col Glenn Bailey, Lt Col Jack K. Kloeber, Jr., Lt Col James T. Moore, and Capt Christopher Burke to the operational analysis process. They joined the Analysis team as expert evaluators of the futuristic systems considered by the *Air Force 2025* study. They were chosen to participate because each had a unique career background that was deemed essential to any attempt at an objective assessment of the futuristic systems considered by *Air Force 2025*. Their contribution was immeasurable.

Executive Summary

In the summer of 1995 the Air Force chief of staff tasked Air University to conduct a year-long study, ***Air Force 2025***, to "generate ideas and concepts on the capabilities the United States will require to possess the dominant air and space forces in the future, detail ... new or high-leverage concepts for employing air and space power, detail ... the technologies required to enable the capabilities envisioned." To support this goal an operational analysis was conducted to identify high-value system concepts and their enabling technologies in a way that was objective, traceable, and robust. This analysis determined which of the ***Air Force 2025*** system concepts showed the greatest potential for enhancing future air and space capabilities and which of their embedded technologies have the highest leverage in making the high-value system concepts a reality.

A model, ***Foundations 2025***, which reflected the overall values held by the ***Air Force 2025*** participants was developed to quantify and compare different system concepts' contributions to future air and space capabilities. ***Foundations 2025*** is distinguished by the large number of system concepts that can be analyzed, the 30-year focus into the future, and the fact it was developed through a bottoms-up approach. ***Foundations 2025*** offers a potential new framework for future air and space doctrine and can be easily modified (broken into three separate models: *awareness*, *reach*, and *power*) by Air Force major commands for use in their mission area analysis process. Thus, the model presented is an aid to current and future senior decision makers concerned with the employment of air and space power.

The ***Air Force 2025*** study produced a number of excellent system concepts for employing air and space power in the future. Analysis of the highest value system concepts indicated that the effort to occupy the "high ground" of the future will require air and space forces to possess increased *awareness* and to control the medium of space. The five highest value system concepts were:

- Global information management system
- Sanctuary base
- Global surveillance, reconnaissance, and targeting system
- Global area strike system
- Uninhabited combat air vehicle

The following six system concepts scored below the top five but were clearly ahead of the others:

- Space-based high energy laser
- Solar-powered high energy laser
- Reconnaissance UAV
- Attack microbots
- Piloted SSTO TAV
- Uninhabited air-launched TAV

These conclusions regarding the rankings of the system concepts were not affected by any reasonable changes of the weighting scheme in the Value Model.

The study also included an assessment of the enabling technologies on which the system concepts depend. The analysis explicitly took into account the number of system concepts each technology supported, the degree to which each system concept depended on it, and the importance of the system concept. Six high-leverage technologies stood out because they are important to a large number of high-value system concepts:

- Data fusion
- Power systems
- Micromechanical devices
- Advanced materials
- High energy propellants
- High performance computing

The major surprise among these results was the importance of continued breakthroughs in the area of power systems. Other moderate-leverage technologies were also important but contributed to only three or four of the high-value system concepts:

- High energy laser systems
- Artificial intelligence
- Optics
- Aerospace structures
- Image processing
- Communications

Advances in these areas show promise to open the way to air and space systems that would dramatically improve the effectiveness of air and space power employment to achieve the US military objectives.

Chapter 1

Introduction

The long range planning process in our Air Force is broken. If we are going to be relevant in the future, we've got to somehow break free of the evolutionary nature of the planning process.

—Gen Ronald R. Fogleman

With these few words, the chief of staff of the Air Force, Gen Ronald R. Fogleman, challenged the participants of the ***Air Force 2025*** study to generate ideas and concepts on the capabilities the United States will require to dominate air and space forces in the future. When General Fogleman assigned the responsibility for ***Air Force 2025*** to Air University, he directed that the final product be a collection of white papers detailing findings regarding air and space capabilities required for future warfare, new or high-leverage concepts for employing air and space power, and the technologies required to enable the required capabilities.[1]

Meeting the Challenge

In response to General Fogleman's tasking, Air University devised a four-phase study process to stimulate creativity, generate ideas, and evaluate figure 1-1.

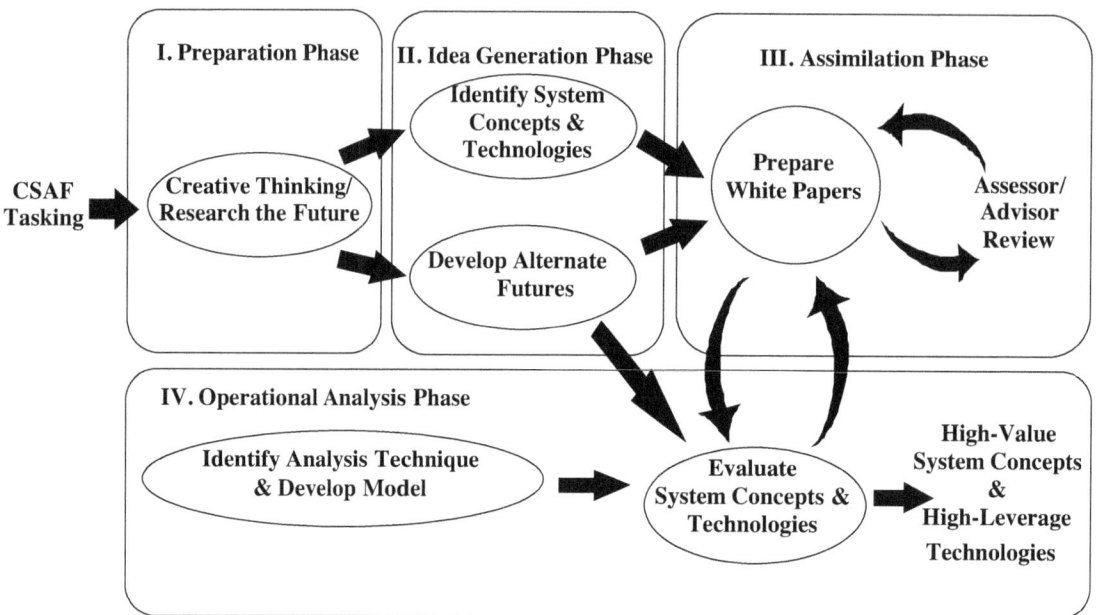

Figure 1-1. *Air Force 2025* **Study Process**

The Preparation phase exposed, participants to a wide variety of creative thinking and problem-solving concepts. This phase laid the groundwork for the Idea Generation phase in which the participants developed plausible alternative futures as well as future system concepts and technologies. Inputs for the Idea Generation phase were gathered from a worldwide data call which produced more than 1,000 submissions.

In the Assimilation phase, the participants were organized into specific writing teams based on operational experience. Each team was given a particular area to consider and concentrate their research (table 4). After postulating the required capabilities of the future Air Force, each team developed system concepts and technologies from the Idea Generation phase to satisfy these future requirements.

This phase produced a large number of system concepts which were described in varying levels of detail, which provided widely different kinds of operational capabilities, and which depended on different levels of advancements in different areas beyond current technology. Clearly, not all of these system concepts could be developed, nor could all of the technologies be aggressively pursued. The study needed to prioritize the relative importance of both future system concepts and their enabling technologies.

An operational analysis was conducted with the other three phases to aid in this prioritization. Its purpose was to evaluate system concepts and technologies developed in the white papers; specifically, it had three objectives:

1. To assess the potential operational utility of future air and space system concepts.
2. Identify the high leverage technologies required by those system concepts, and
3. Provide an objective, traceable, and robust analysis.

Figure 1-2 shows the methodology used for the operational analysis. The left side of the column reflects the evaluation of system concepts for operational utility, while the right column identifies and evaluates the underlying high-leverage technologies.

For the system concept evaluation, this model used a hierarchy of objectives, functions, and tasks, which represented the objectives necessary to underwrite air and space power in the next 30 years. For the technology evaluation, the framework was a logical structuring of technology areas that were mutually exclusive and collectively exhaustive. These hierarchies provided the desired characteristics of objectivity and traceability. The desired robustness quality was assured by performing a sensitivity analysis at the conclusion of the system concept and technology scoring. Specifically, the sensitivity analysis was conducted across a number of plausible alternate futures (see "Alternate Futures section of chap. 2).

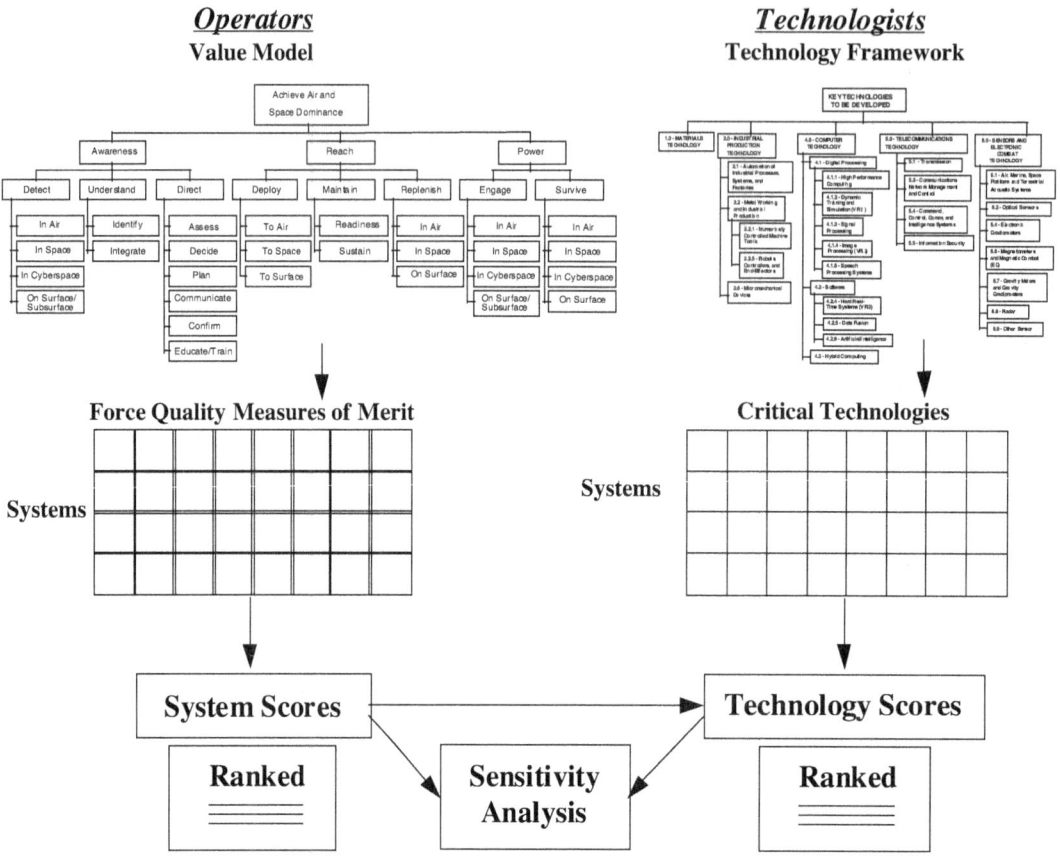

Figure 1-2. Operational Analysis Methodology

Overview of Report

This report describes in detail the operational analysis conducted as a part of the *2025* study and introduces *Foundations 2025*, a value model for evaluating future air and space capabilities. Chapter 2 discusses the methodology used to develop the *Foundations 2025* value model, the technology framework, and how each was used to evaluate the *2025* systems and underlying technologies[2]. The results of the analysis are presented in chap. 3 and conclusions are summarized in chap. 4. Several appendices are included and provide the supporting data for Chap. 3 and 4.

[1] Message from Gen Ronal R. Fogleman to Air University, 23 December 1994.

[2] From this point forward we will use the term *system* when referring to the system concepts. The authors recognize that *system* carries the connotation of existing hardware, but it is less cumbersome, and all of the systems scored here are futuristic.

Chapter 2

Methodology

This chapter outlines the *Air Force 2025* methods used to evaluate systems and technologies. It covers the development of the value model to score the systems, the system identification process, the system scoring procedures, the technology identification procedures, and the technology scoring procedures and ends with an evaluation of which sector (public or commercial) will develop the future technologies.

Challenges

A primary goal of the *Air Force 2025* operational analysis (OA) was to identify the *Air Force 2025* systems that offer the greatest potential to support future air and space operations. To meet this goal, the Analysis team's challenge was to develop a methodology that satisfied a diverse set of criteria. First, the *Air Force 2025* OA needed to be compatible with the Air University academic calendar year. It also needed to be capable of quick implementation after the Air Command and Staff College and Air War College students completed their white papers, which contained conceptual descriptions of the systems.

Second, because *Air Force 2025* was a study about 30 years into the future, the system descriptions in the white papers lacked engineering detail. Therefore, the OA methodology had to use human judgment about operational capability and key enabling technologies.

Third, while the values of the current senior leadership of the Air Force are well documented in strategies, policies, and directives, it is far more difficult to predict what will be important to future leaders.

Fourth, to prevent one set of views or interests from unduly influencing the results, the evaluation methodology had to be free of institutional bias. The methodology should neither unfairly favor nor unduly penalize any potential *Air Force 2025* systems.

Fifth, the results had to be traceable, since the *Air Force 2025* system evaluation results would be subject to scrutiny. The Analysis team members would need to explain for any given system or technology how and why it was scored. The study participants and Air Force senior leadership would be far more likely to accept the results if they could understand clearly how the systems were evaluated.

Sixth, the OA methodology had to be robust to apply across a wide range of potential future environments postulated by the *Air Force 2025* Alternate Futures team. Each future described a different political, technological, and social environment (see the "Alternate Futures" section). The OA methodology had to be able to capture different priorities assigned to air and space functions and tasks in these alternate futures.

The *Air Force 2025* Operational Analysis

The Analysis team developed the four-step approach to satisfy the requirements previously described figure 2-1. This section contains a very brief overview of the *Air Force 2025* OA process; the following sections describe each of the four steps of the process in detail. First, the Analysis team had to choose the most appropriate analysis methodology for the OA. After considering the advantages and disadvantages of many possibilities, the team selected an approach known as Value-Focused Thinking (VFT).[1] Second, to implement VFT, the team had to find a *value model* to quantify and compare the contributions of the systems proposed by *Air Force 2025*; therefore, the next step was a comprehensive search for a suitable existing value model. Third, when no suitable alternative was found, the Analysis team developed a new value model, **Foundations 2025**. The fourth step used the value model to evaluate *Air Force 2025* systems and then the technologies that will be required to develop these systems.

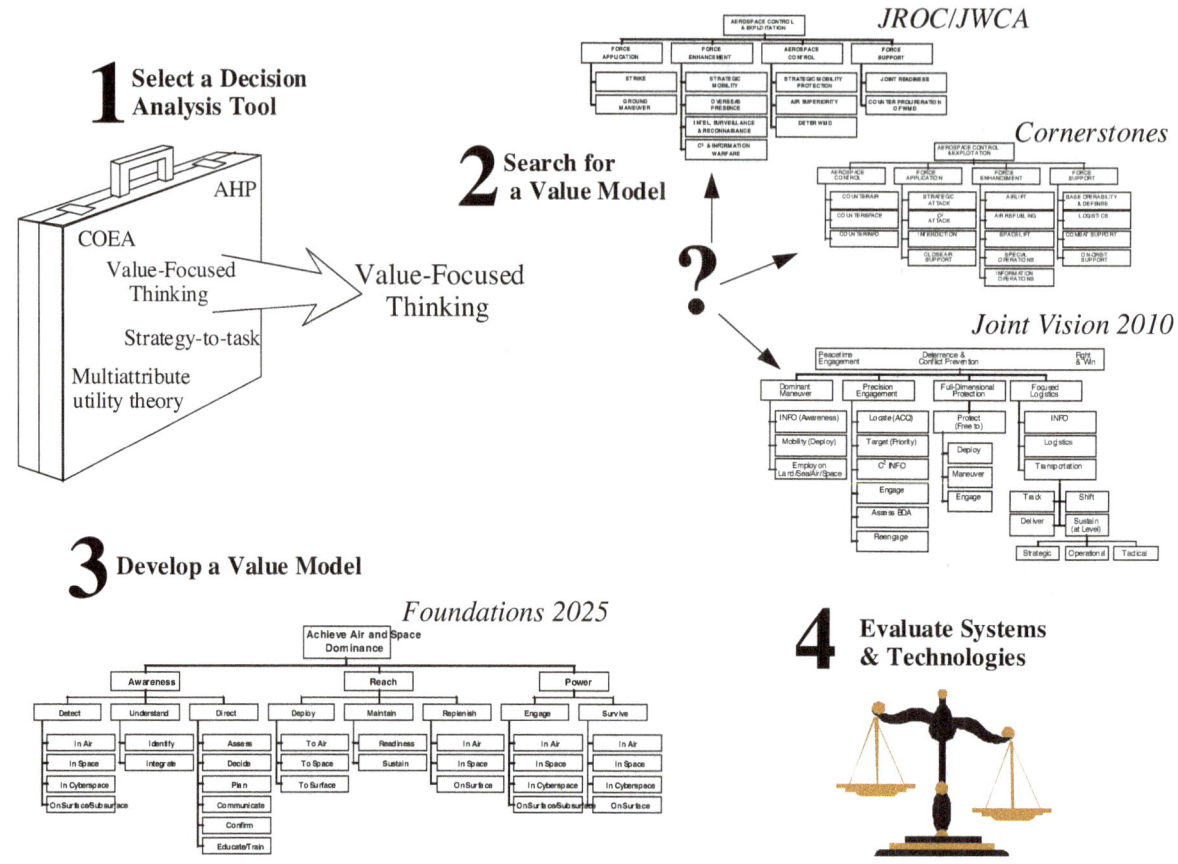

Figure 2-1. *Air Force 2025* OA Process

Selecting a Decision Analysis Tool

The first step in the *Air Force 2025* OA was to find the analytic approach that best satisfied the study requirements. The challenge facing the Analysis team could be stated simply: Given a large number of ideas for future systems and technologies, how should the Air Force evaluate and compare the relative "worth" of competing alternatives ?

Comparing Analysis Tools

Each analysis approach has particular strengths and weaknesses; therefore, the Analysis team examined them in relation to the challenges of the *Air Force 2025* study discussed previously. The team considered the following analysis techniques:

- "most-to-least dear" with no criteria
- qualitative comparison with criteria
- simple quantitative comparison matrix
- value-focused thinking
- analytical hierarchy process
- strategy-to-task
- futures-to-strategy-to-task
- common operational objectives of the armed forces
- cost and operational effectiveness analysis

These choices are listed in order of increasing depth of the analysis. In selecting an analysis technique for this task, the most important tradeoff is between the level of engineering detail and the available time. The first few methodologies require the least amount of detail concerning the evaluated systems, while the last few demand considerable engineering detail. At one extreme ("most-to-least dear"), a group of experts can review the alternatives and can rank them subjectively. At the other extreme, a full cost and operational effectiveness analysis can be done, as is usually done before starting the development phase of any new major weapon system acquisition.

The first approach is "most-to-least dear" and has no criteria. This technique is sometimes referred to as the "bunch-of-smart-people-around-a-table" approach. This method can be accomplished quickly; however, it is difficult to replicate. This approach is not robust, and it lacks objectivity and traceability. The second approach, qualitative comparison with criteria, is slightly more rigorous, but the subjective rationale again lacks objectivity and traceability. The next approach, a simple quantitative comparison matrix, provides more rigor, but with the large number of systems across all air and space capabilities, the team felt a comparison matrix approach would not adequately distinguish between competing systems.

The fourth approach, value-focused thinking, was the technique used in the Air University ***SPACECAST 2020*** 1994.[2] Instead of highly detailed system data, VFT employs user-defined scoring functions to compare the performance of alternate systems. Systems that perform well on these scoring functions receive high-value scores. The VFT process is traceable because system evaluations are based on defined scoring functions. This approach capitalizes on a strength of the ***Air Force 2025*** process — lots of expert judgment.

The fifth technique, the analytical hierarchy process (AHP), ranks alternatives, but its several drawbacks limited its usefulness for this study. AHP requires pairwise comparisons of systems between to each criterion, as well as among criteria. The large number of systems made such comparisons cumbersome.

The next three techniques—strategy-to-task, futures-to-strategy-to-task, and common operational objectives of the armed forces—require an operating concept and a known strategy. The *Air Force 2025* study, by its nature, moves beyond the point where well defined operating concepts support a known strategy.

At the end of the spectrum is the cost/operational effectiveness analysis (COEA) technique. This approach is the most rigorous and time consuming because it demands detailed engineering data which does not exist. Furthermore, COEAs typically take several months or sometimes years to complete.

The Best Approach for *Air Force 2025*

After considering the advantages and disadvantages of the various approaches, the Analysis team believed VFT offered the best compromise for satisfying the OA requirements. VFT was particularly suited for structuring the subjective judgments required to evaluate the systems. It also allowed the OA to be completed in the limited time available and, because VFT was used in the *SPACECAST 2020* study, it was well understood and accepted by the Air University senior leadership. In addition, once a value framework was built using VFT, it was easy to assess systems across several alternate futures. Finally, the VFT methodology enables the OA to be objective, traceable, and robust.

Value-Focused Thinking

VFT begins by identifying the decision maker's values a hierarchy of objectives. Top-level objectives describe aspirations that are most important to the decision maker. Objectives are decomposed until desired force qualities can be specified and measured. Weights are assigned to signify the relative importance of objectives at every level.

In the VFT methodology, we use several key terms—*value, objectives, functions, tasks, subtasks, force qualities, measures of merit, scoring functions, value model,* and *weights*.

Value

The most important concept in VFT is *value*. Ralph L. Keeney says, "Values are what we care about. [Values] should be the driving force for our decision-making."[3] Instead, the fundamental precept of VFT is that values are principles used for evaluation.[4]

Keeney asserts that thinking about values is constraint-free thinking.[5] By not limiting oneself to pre-etermined alternatives in the search for solutions, innovative answers can be developed. This philosophy complemented the creative thinking exercises of *Air Force 2025*.

Objectives, Functions, Tasks, and Subtasks

In VFT, values are made explicit with objectives, and a hierarchy of objectives is constructed that supports the decision maker's values.[6] Specific, lower level objectives support the general, overarching objectives. The Analysis team used the terms *objective*, *functions*, *tasks*, and *subtasks* to designate the tiers in the hierarchy, from highest to lowest, respectively.

Force Qualities

In VFT terminology, a force quality defines a desired attribute of a system to achieve a subtask. For example, if the subtask is to "identify," a corresponding force quality might be "accurate." According to Keeney that "[force qualities] should be measurable, operational, and understandable."[7]

Measures of Merit and Scoring Functions

Each force quality has a measure of merit that is the metric used to gauge system performance. Each measure of merit has a range of outcomes, from worst to best. To continue with the previous example, if the subtask is "identify" and the force quality is "accurate," then a measure of merit could be "percent of correct identifications."

VFT scoring functions provide a quantitative means for measuring the relative system performance for each measure of merit. For example, if the measure of merit is percent of correct identifications, the corresponding scoring function might convert a system performance of "83 percent correct identifications" into a score of 92.

Because scoring functions operationalize the measures of merit, they are the building blocks for assessing system performance. The domains (horizontal axes) of the scoring functions are the measures of merit; the ranges (vertical axes) are the corresponding value scores. A scoring function's domain may be quantitative or qualitative, but its range must be quantitative. While domains are different, ranges must be the same. Typical ranges are [0,1], [0,10], or [0,100]. The development of these functions is a significant analytical task, since analysts must discern the decision makers' values for the full range of the measure of merit for each force quality.

Many types of scoring functions exist; the relationship between subtasks and force qualities dictates the appropriate form. One of the simplest scoring functions is the linear scoring function figure 2-2.a. If a decision maker chooses a linear function, the underlying implication is that each incremental increase in performance is valued just as much as the preceding increment.

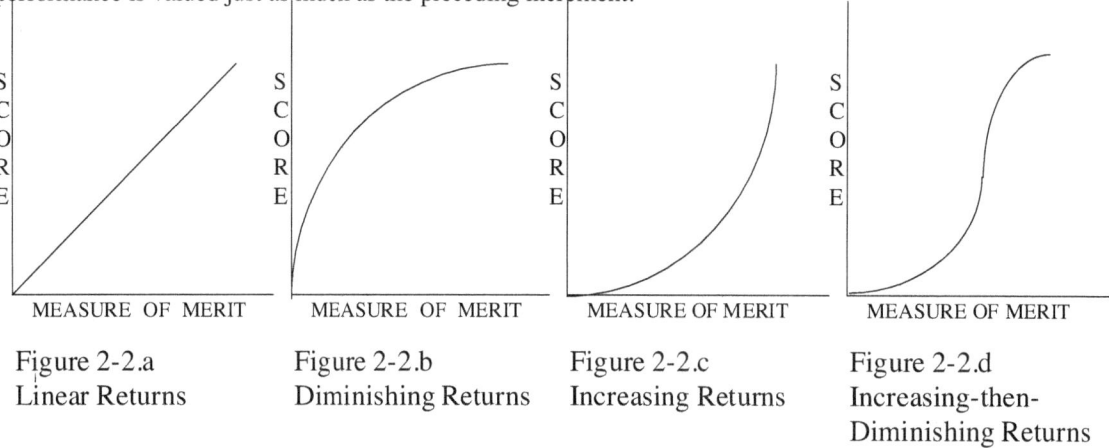

Figure 2-2.a
Linear Returns

Figure 2-2.b
Diminishing Returns

Figure 2-2.c
Increasing Returns

Figure 2-2.d
Increasing-then-
Diminishing Returns

Figure 2-2. Sample Scoring Functions

The linear scoring function has the advantages of easy construction and computation; however, decision makers frequently do not assign value linearly. For example, the value placed on system performance often experiences diminishing marginal returns (fig. 2-2.b). In other words, each incremental increase in performance is valued less than the preceding increment. Another way of explaining diminishing returns is to say that some level of performance is "good enough."

Another phenomenon is that of increasing marginal returns (fig. 2-2.c). Here each incremental increase in performance is worth more than the preceding increment. Under these circumstances, the decision maker admits there is a certain threshold of performance that must be met before the system has substantive value.

Another popular scoring function is the S-curve, which initially reflects increasing returns, then switches to diminishing returns (fig. 2-2.d). The preceding four examples are the most common scoring functions; however, as long as the objective is to maximize value, any monotonically increasing (i.e., never decreasing) representation is possible. R. T. Clemen,[8] Keeney,[9] or Wayne Winston[10] techniques for deriving scoring functions.

Value Model

A value model is the hierarchical representation of objectives, functions, tasks, subtasks, force qualities, measures of merit, and scoring functions. ***Foundations 2025*** was the value model developed for ***Air Force 2025***. A value model, also called a value tree by some authors, is a branching structure, with the most fundamental decision maker objectives at the top. Keeney uses the term *fundamental objectives hierarchy*,[11] and states, "The higher level objective is defined by the set of lower-level objectives directly under it in the hierarchy."[12] In other words, the lower level objectives completely specify their higher level objective.

Clemen describes five specific characteristics of a value model:

1. It should be complete, encompassing all important facets of the decision.
2. It "should be as small as possible."[13]
3. The force qualities should allow straightforward measurement.
4. Objectives should only appear once in the tree.
5. The decision maker should be able to think about and treat the branches of the tree separately.[14]

Combining the first, fourth, and fifth characteristics above yields two properties—the objectives must be mutually exclusive (only appear once and can be treated separately) and collectively exhaustive (encompass all that the decision maker values).

Analysts work with decision makers to develop value models. Sometimes candidate value models already exist, at least partially, within an organization. In other cases, a new value model must be constructed. Keeney provides further background and techniques for developing value models.[15]

Figure 2-3 shows a notional value model. Under the objective, there are three functions (first level of subordinate objectives). Under functions are corresponding tasks (second level of subordinate objectives). In this example, function 2 has two tasks (A and B). Tasks can also have subtasks(not included in this example), and tasks and subtasks can have one or more force qualities that characterize success at this task. Task A has three force qualities and Task B has one force quality (fig. 2-3). For each force quality, one or more measures of merit must be identified, each of which has a corresponding *scoring function*.

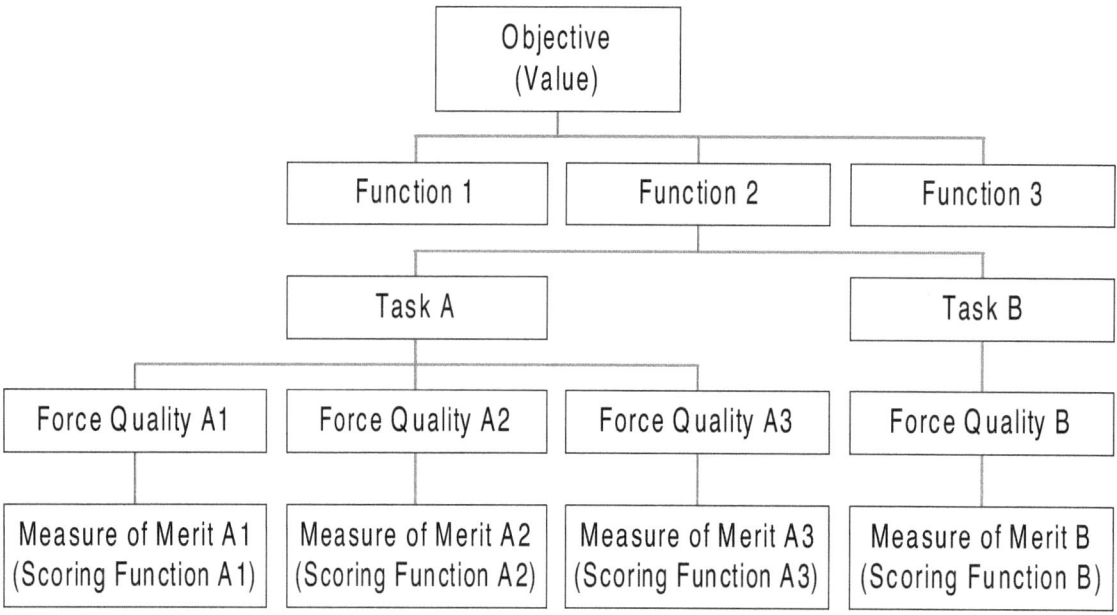

Figure 2-3. Notional Value Model

Weights

After the hierarchical structure of the value model is complete, the decision maker must determine the relative importance of the functions, tasks, force qualities, and measures of merit. Numerical weights are assigned across each tier of the value model. These weights must satisfy certain mathematical requirements. Weights must be between 0 and 1, and for each function or task, the weights of the immediately subordinate functions, tasks, subtasks, and force qualities must sum to 1.00. Figure 2-4 shows weights applied to the earlier notional value model. Note that the weights of the three functions sum to 1.00, as do the weights of the two tasks under function 2 and the three force qualities under task A. Note also that when a subordinate level contains only one box, its weight is necessarily 1.00.

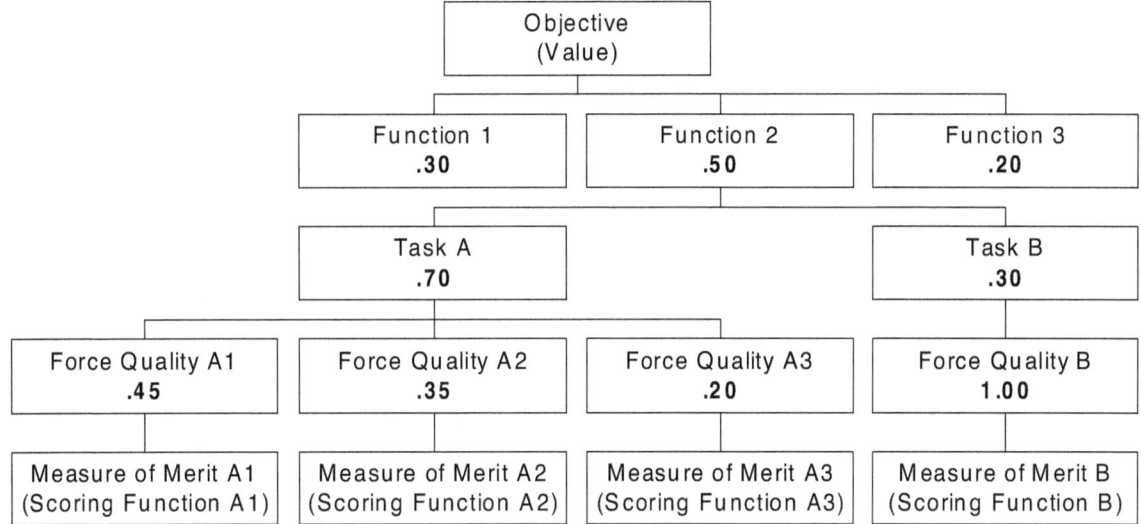

Figure 2-4. Weights Applied to Notional Value Model

Applying the Value Model

After the value model is completed and weights has been assigned, the evaluation of systems can be accomplished. The computations are straightforward. First, for system, raw scores are initially computed by using the applicable scoring functions. If a particular measure of merit does not apply to a given system, the raw score is zero. Next, weighted scores are calculated by multiplying the raw scores by the product of all the weights in the branch that leads (bottom-up) from the measure of merit to the highest level objective. Finally, a total system score is found by summing the weighted scores. The system with the highest score is the "best".

Using the earlier notional example (fig. 2-4), assume system X has raw scores of 37, 62, 18, and 83 for measures of merit A1, A2, A3, and B, respectively. The resulting score for function 2, S(function 2), is

$$\begin{aligned} S(\text{function 2}) &= \{[(37)(0.45)(0.7)(0.5) + (62)(0.35)(0.7)(0.5) + (18)(0.2)(0.7)(0.5)] + (83)(1.0)(0.3)(0.5)\} \\ &= 5.82 + 7.59 + 1.26 + 12.45 \\ &= 27.12 \end{aligned}$$

Similar operations are performed for every branch of the value model, then they are summed to compute the total score for system X.

The Search for the *Air Force 2025* Value Model

After the Analysis team selected VFT, the next step was to either select an existing value model or develop a new one. Identifying a current model proved to be a daunting task because of the scope of the study and the focus on the far future. The participants ranged across all of the military services and included allies, civilians, government officials, and industry. Any potential model also had to satisfy Clemen's five criteria.[16]

The Analysis team initially searched for a national level strategic document that identified priorities for future air and space forces. The following sources were investigated:

- A National Security Strategy of Engagement and Enlargement
- National Military Strategy of the United States of America
- Defense Planning Guidance
- Joint Requirements Oversight Council (JROC)/Joint Warfighting Capabilities Assessment (JWCA) categories
- Global Presence and Global Reach, Global Power
- Common operational objectives of the armed forces
- Draft AFDD-1
- Joint Vision 2010
- Cornerstones of Information Warfare

A National Security Strategy of Engagement and Enlargement[17] represents the views of the President. It provides a high-level strategic view of the national defense policy but does not support or provide any definitive characteristics that could form the basis of a value model. The *National Military Strategy of the United States of America* (NMS), by the chairman Joint Chiefs of Staff (CJCS), also has a grand strategy perspective.[18] The NMS contains a framework, but the task breakdowns have too much overlap. The *Defense Planning Guidance*[19] is focused on the near- and midterm budget cycle; therefore, it does not support the visionary requirements of ***Air Force 2025***.

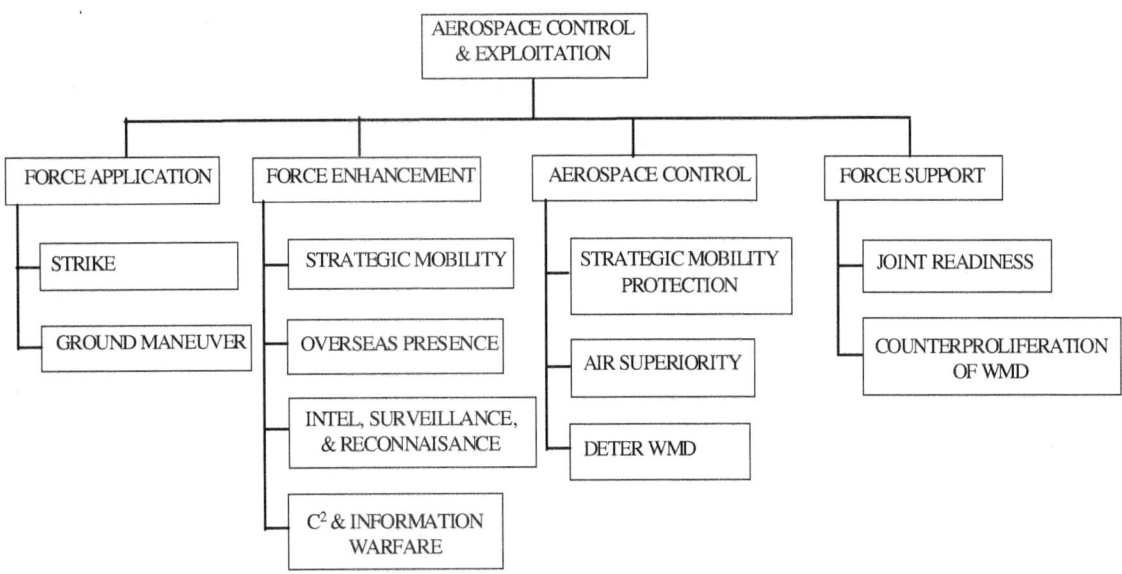

Figure 2-5. Value Model Based on JROC/JWCA Categories

The former vice CJCS, Adm William A. Owens, produced a set of JROC/JWCA categories to support the planning and analysis integration of the service program objective memoranda (fig. 2-5).

While the JROC/JWCA framework was designed to be visionary, it also has considerable overlap. For example, the tasks of to "deter weapons of mass destruction (WMD)" and "counter proliferation of WMD" are not mutually exclusive, nor are the tasks of "ground maneuver," "strategic mobility," and "strategic mobility protection."

The secretary of the Air Force published the *Global Reach, Global Power*[20] and *Global Presence*[21] series of documents to promulgate the high-level vision of the Air Force of the future. But like the president's national security strategy, these documents do not define key functions required to create a hierarchical value framework.

Another interesting approach, common operational objectives of the armed forces, is based on the work of Lt Gen USAF, Glenn A. Kent (Retired).[22] This document contains a strategy-to-task focus that could be suitable; however, like the defense planning guidance, it focuses on midterm planning years. This approach has two other qualities that made it unsuitable for *Air Force 2025*: it requires more engineering definition of the systems and concepts being evaluated than is available, and it assumes a working concept of operations. Moreover, it assumes a known future strategy.

The Analysis team next looked to doctrinal publications as a potential source for a value model framework because doctrine is specifically written to provide guidance on how to think about air and space capabilities without being tied to a particular time period. The Air Force is currently writing a new doctrinal pamphlet, *Air Force Doctrine Document: Air Force Basic Doctrine - 1* (AFDD-1).[23]

Figure 2-6 shows a candidate value model based on the first draft of AFDD-1. This framework has several shortcomings. First, the tasks of "strategic attack," "interdiction," and "close air support" contain considerable overlap. Next, "information operations" is not integrated into the framework but rather is represented as a series of traditional functional tasks such as "C^4", "intelligence," and "weather service." The goal of the *Air Force 2025* study was to propose and evaluate systems that are employed to dominate air and space; it was not concerned with how we are currently organized to perform traditional functions.

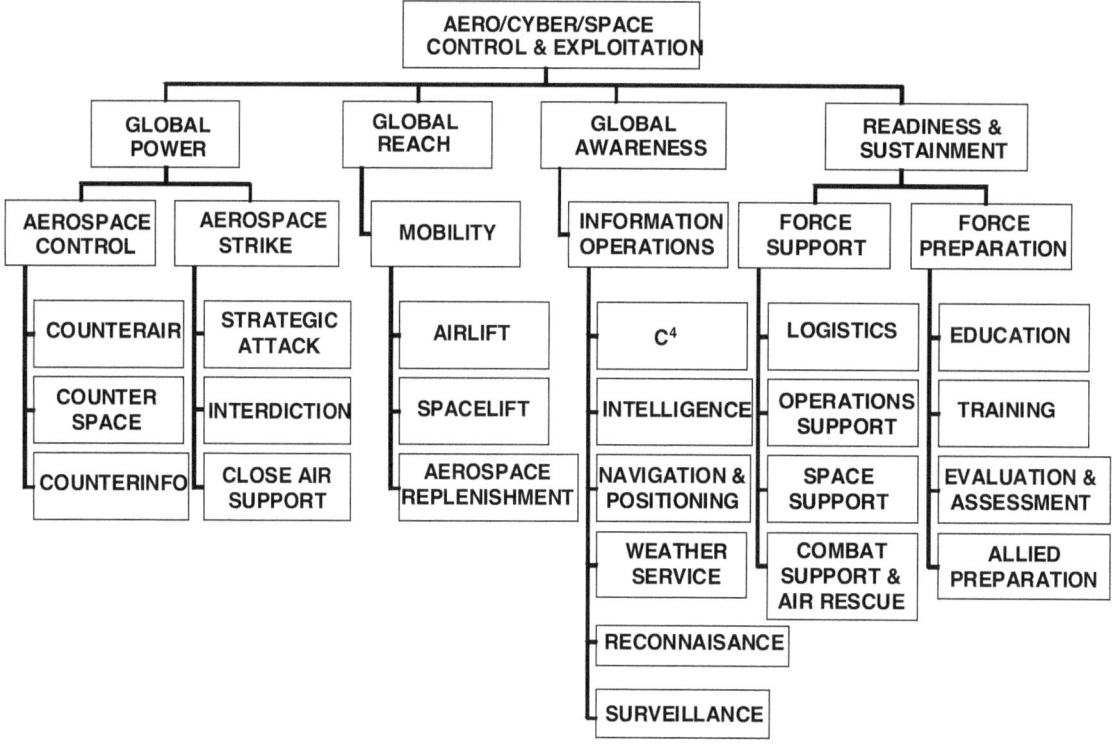

Figure 2-6. Value Model Based on AFDD-1

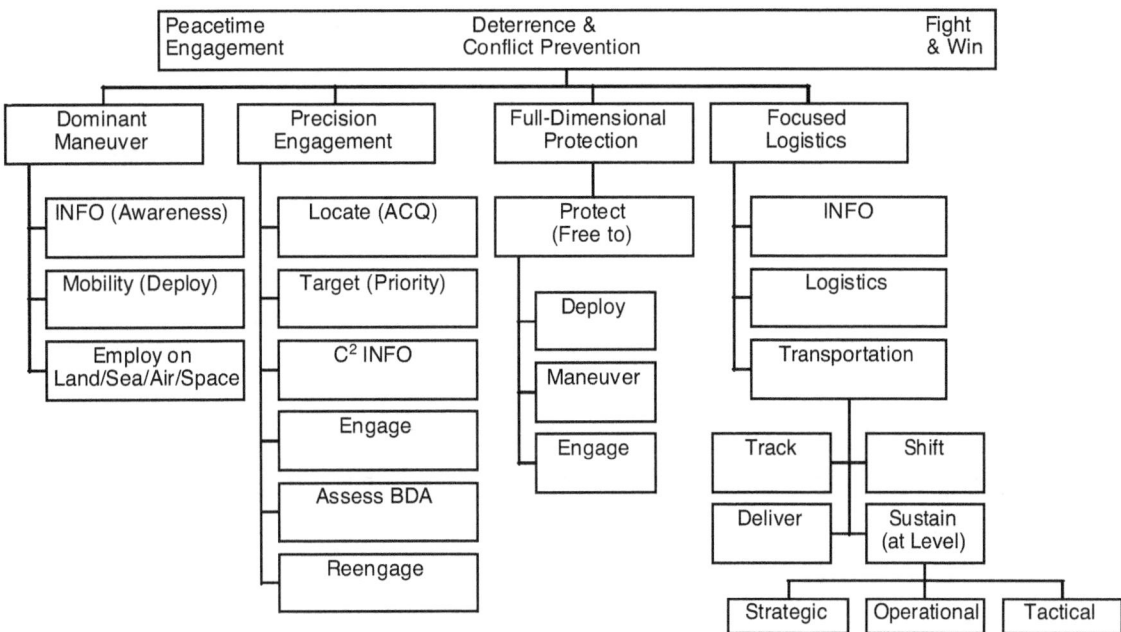

Figure 2-7. Value Model Based on *Joint Vision 2010*

Another doctrinally based candidate for the ***Air Force 2025*** value framework was based on a document from the Joint Staff, *Joint Vision 2010* (fig. 2-7).[24] As its title states, *Joint Vision 2010* focuses on the future 15 years hence, not 30 years. In many cases, the acquisition process for the systems of 2010 has started already, so this model is not sufficiently visionary. On the other hand, *Joint Vision 2010* contains a number of interesting departures from current service doctrines, especially in the focus on precision engagement and the vision for information and maneuver dominance. However, the model contains significant overlap in the tasks of "mobility (deploy)," "deploy," "maneuver," and "transportation." Because of the overlap, and since it did not look far enough into the future, the Analysis team rejected the *Joint Vision 2010* model.

The existing model with the closest fit to the ***Air Force 2025*** requirements was in the document *Cornerstones of Information Warfare*, released by the Chief of Staff and the secretary of the Air Force in late 1995.[25] *Cornerstones* was written "to provide a sound and widely accepted basis from which we can adapt Air Force doctrine to the Information Age."

Figure 2-8 shows the *Cornerstones* hierarchical framework. *Cornerstones* seemed to accommodate most of the tasks and functions envisioned in ***Air Force 2025***. It addresses information warfare in a realistic and rational manner and is written with time-honored terms that are easily understood by the study participants. Except in the area of information operations, it has the only mutually exclusive and collectively exhaustive structure of any of the existing frameworks and, therefore, provides a sound mathematical basis for analysis. While an excellent document in many respects (its view of information warfare as another technique for mission accomplishment is noteworthy), *Cornerstones* continues to maintain a number of traditional "stovepipe" functions. As a result, the institutional biases inherent in the model could have stifled the creative process of *Air Force 2025*.

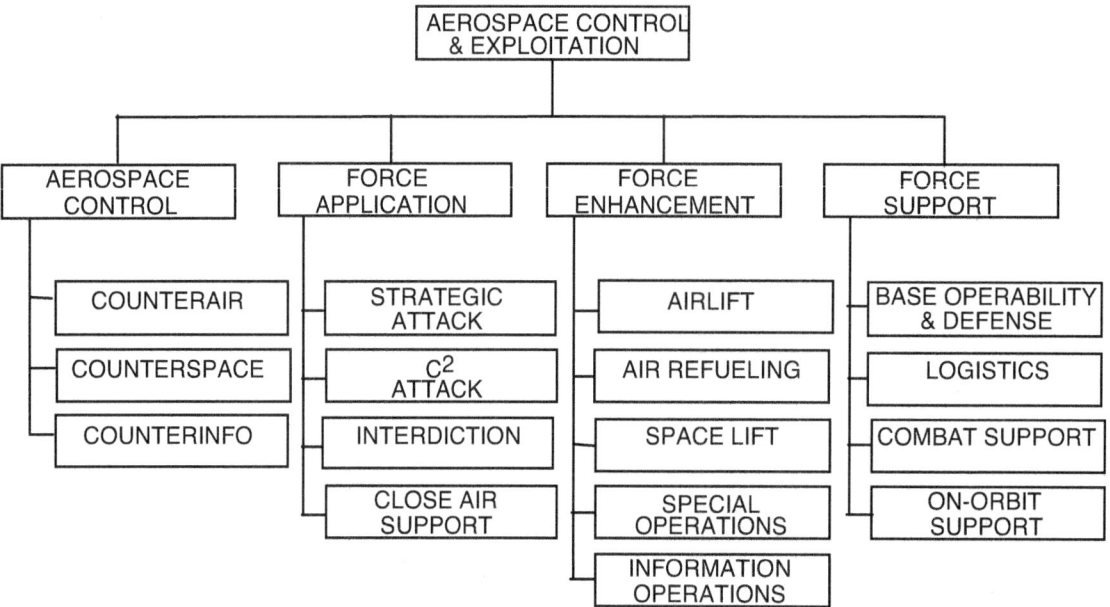

Figure 2-8. Value Model Based on *Cornerstones*

After reviewing the full range of potential value models, the senior leadership of *Air Force 2025* determined that none of the models met the requirements of *Air Force 2025*. Each model was grounded in near- or midterm thinking, and none seemed to think "outside of the box" about new ways to employ air and space forces in the far future. Furthermore, each contains traditional biases focusing on how the Air Force is organized, while *Air Force 2025* addresses the dominant employment of air and space forces in the year 2025 and beyond. The only solution was for the Analysis team to develop a new framework to capture the visionary thinking that took place during the study.

Developing the *Air Force 2025* Value Model—*Foundations 2025*

Developing the *Air Force 2025* value model was a key part of the analysis process. The work began early in *Air Force 2025*, and it continued for the duration of the study. The final value model, ***Foundations 2025***, was so named because it provided the basis for the *Air Force 2025* operational analysis.

This section traces the evolution of the ***Foundations 2025*** value model. The process of developing the model began by identifying the model's overarching objective. Next, the Analysis team built tasks and functions from the bottom up, an atypical approach to building value models. When the task-function-objective framework was complete, the team completed the value model by creating force qualities, measures of merit, and scoring functions.

Objective

Before making any progress toward developing a value model, the Analysis team needed a clear statement of the objective. As stated in the "Introduction," General Fogleman tasked the *Air Force 2025* participants to generate ideas and concepts on the capabilities the United States will require to dominate air and space in the future. This statement was translated into the overarching objective, "Achieve Air and Space Dominance," that became the top tier of ***Foundations 2025***.

A Bottom-Up Approach

With this overarching objective defined, the Analysis team could start specifying subtasks, tasks, and functions. Early on, the team departed from the usual approach to constructing a value model. Conventional value models are built in a top-down fashion; each level of the model hierarchy is derived from the next higher level. In contrast to the top-down method, a bottom-up approach makes no *a priori* assumptions and does not establish preconditions. The bottom-up approach should result in less institutional bias.

Tasks and Subtasks

The first step in the bottom-up process was to meet with each of the *Air Force 2025* white paper writing teams to gather their specific notions of the required *tasks* and *subtasks* of future air and space forces. (At this stage, the terms interchangeable, so for now, both are called tasks.) The teams held diverse views on task specification, but the ideas could all be characterized as some form of action. As a result, action verbs were used to specify tasks required for the employment of air and space forces in the year 2025.

Figure 2-9 shows the disparate tasks that resulted from brainstorming sessions with *Air Force 2025* teams. In this figure, the superscript after the task shows the letter designation of the team that identified the task (table 4). The final tally was 109 different tasks; the first step in the evolution of the *Foundations 2025* model was complete. The next challenge for the Analysis team was to reduce this collection into a mutually exclusive and collectively exhaustive set. The affinity exercise proved to be an ideal method for this task aggregation.

The Affinity Exercise

An affinity exercise (also called an affinity diagram) is a method designed to allow a team of people to generate a topical list of ideas creatively and subsequently arrange those ideas into natural groupings. Once similar ideas are grouped together, the team assigns names or headers to each collection of ideas.[26] The affinity exercise is a simple yet potentially fruitful process. Besides fostering creativity by each participant, it "encourages non-traditional connections among ideas." Identifying these nontraditional connections among ideas, thus breaking free from contemporary biases, was the aim of the Analysis team. A final benefit of the affinity exercise is that it encourages mutual exclusivity because all similar tasks are grouped together and confined to a single location in the model.

The Analysis team applied this technique by combining the 109 tasks in figure 2-9 into logical groups. The result of this first affinity exercise was a set of 14 mutually exclusive, collectively exhaustive tasks (Figure 2-10). These 14 task groupings were collectively exhaustive because every one of the 109 tasks in figure 2-9 was gathered into one of the 14 groups. The 14 tasks represented by these groups are mutually exclusive because each of the boxes in figure 2-10 represents a fundamentally different activity. The initial tasks grouped inside the boxes, of course, represent fundamentally similar tasks.

Next, the Analysis team assigned a name to each of the groups. The group name came usually, but not always, from one of the initial tasks in the group. The *Foundations 2025* glossary (appendix A) formally defines the group names.

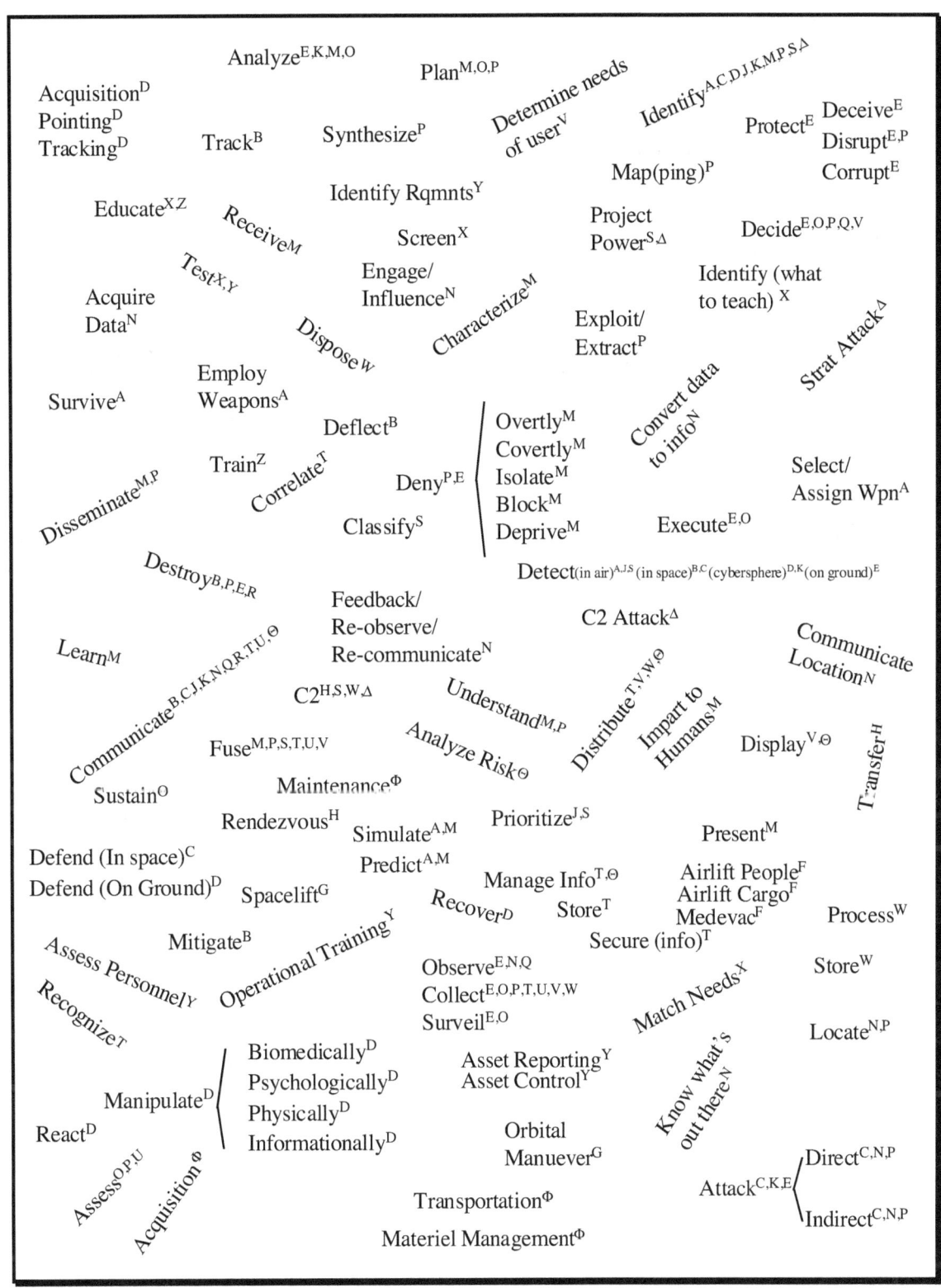

Figure 2-9. *Air Force 2025* Initial Task Compilation

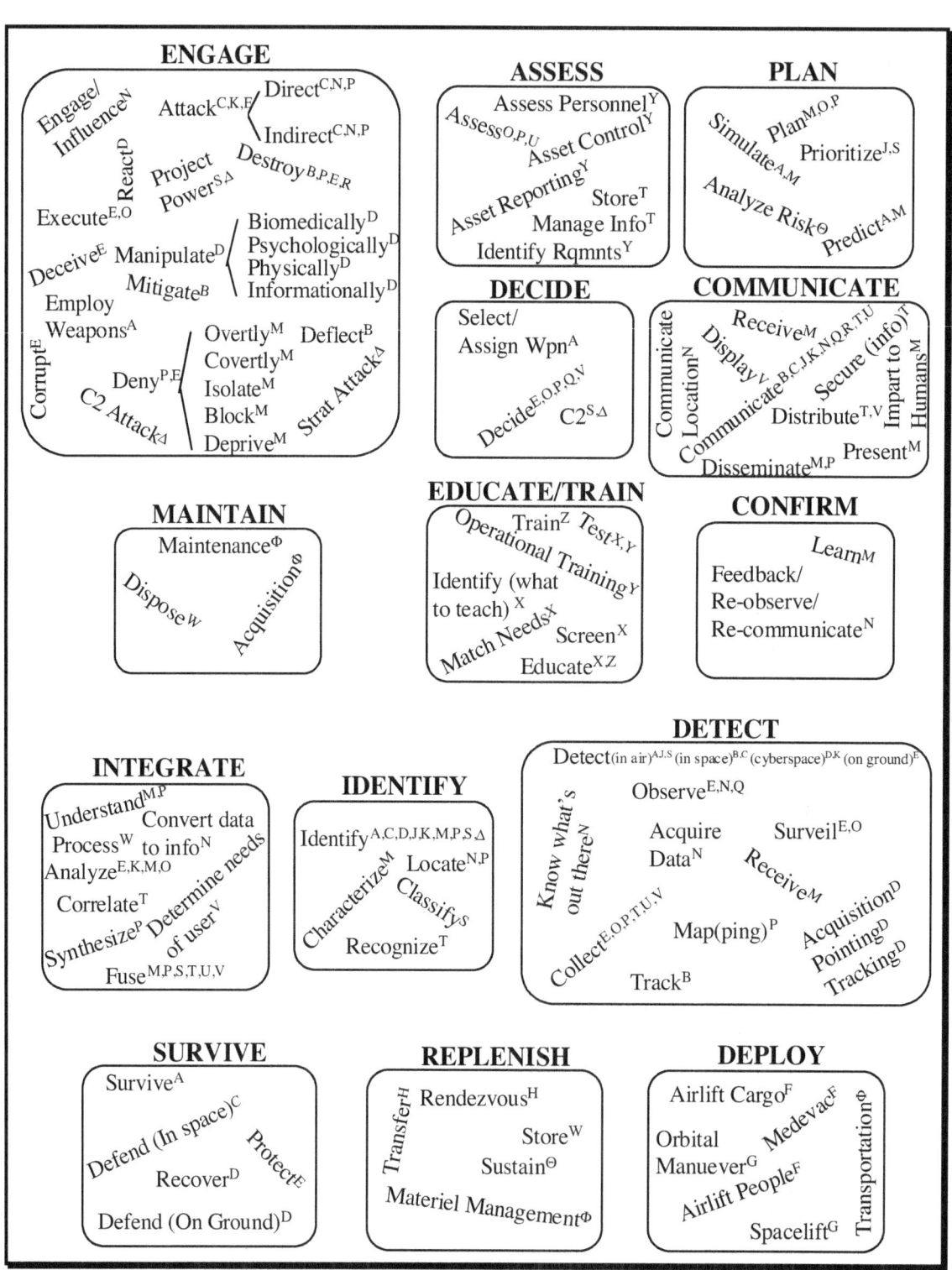

Figure 2-10. First Affinity Grouping of Tasks

Distinguishing Subtasks from Tasks

The second affinity grouping determined which of the 14 tasks in figure 2-10 were truly tasks and which were lower level subtasks. As figure 2-11 shows, six of the 14 tasks were grouped together under a broader task labeled *direct*, and two of the 14 were grouped into a broader task called *understand*. These subtasks were grouped together because, even though they represented different activities, they were closely related. The six tasks under direct—assess, decide, plan, communicate, confirm, and educate/train—became subtasks, as did integrate and identify under understand.

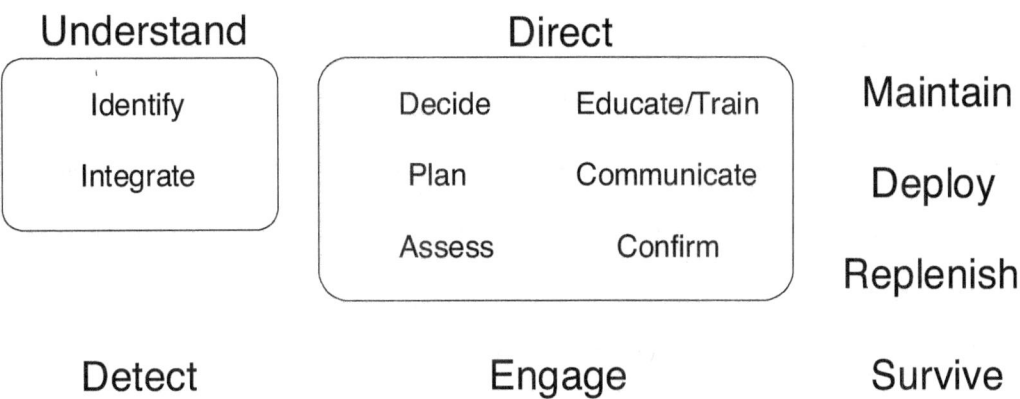

Figure 2-11. Second Affinity Grouping of Tasks

The Influence of the Medium on Certain Tasks

The *Air Force 2025* white paper writing teams felt that some tasks in figure 2-11 were fundamentally different, depending on the medium in which the tasks take place. For example, to detect something in the air is fundamentally different from detecting something on the ground or in cyberspace. Likewise, to engage a target in space is a different task than to engage on the surface. Therefore, the Analysis team had to determine which of the eight tasks were affected by the medium, and, if so, divide those tasks into subtasks to ensure mutual exclusivity.

The four possible media were defined to be air, space, cyberspace, and surface/subsurface; therefore, under the task detect, there are four subtasks: detect (things) in the air, detect in space, detect in cyberspace, and detect on the surface (and subsurface). This same logic applied to the engage, survive, deploy, maintain, and *replenish* tasks. Note that the medium where the target (the thing being detected or engaged) is located, not where the system doing the detecting or engaging is located.

Three tasks in **Foundations 2025** were determined not to be medium specific: understand, direct and maintain. To understand (integrate and identify) what was detected does not depend on where the integrating and identifying occurs. Imagine a fusion center receiving data from sensors in the air, in space, and on the ground. The process of understanding the data passed to the fusion center is not affected by the location of the center. Likewise, the commander's ability to direct, including assess, plan, and decide is not affected by the medium in which these activities occur. Similarly, the ability to maintain is related to readiness and the ability to sustain, but not specifically to a medium.

Two points about the medium-specific subtasks warrant explanation. First, the notions of *deploy to air* and *replenish in air* seemed at first to be far-fetched or unnecessary. However, one of the proposed **Air Force 2025** systems was an airship with the ability to loiter in the air for several days. **Foundations 2025** needed to be sufficiently robust to evaluate such systems and forward thinking in establishing the value of these futuristic tasks.

Second, there were no subtasks in the cyberspace medium for the *deploy, maintain,* and *replenish* tasks. In this paper, cyberspace is defined as "the *virtual* [emphasis added] space of computer memory and networks, telecommunications, and digital media."[27] Because cyberspace is a virtual space, there is no need to move or deploy. The Internet is a good example; being on-line means being engaged.

A medium-specific distinction does not imply that a particular system cannot perform in different media. By dividing some tasks into subtasks based on medium, distinct system capabilities were evaluated based on the phenomenology associated with the medium. In the **Foundations 2025** value model, a satellite that can detect targets on the surface, in the air, and in space received credit for performing three distinct subtasks.

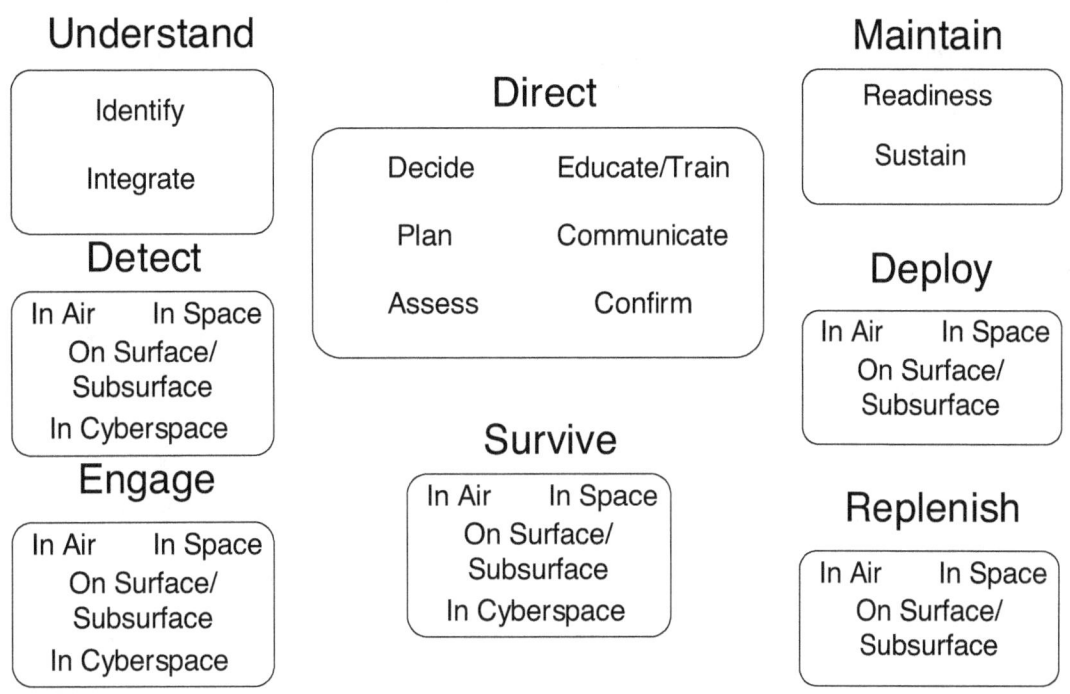

Figure 2-12. Inclusive Grouping of Tasks and Subtasks

Adding medium-specific subtasks increased the number of subtasks in the model from eight to 29. Figure 2-12 shows the complete list of tasks and subtasks, including medium-specific subtasks. The next step in the hierarchical evolution of the value model was to generate functions.

Functions

Functions are the high-level, aggregated tasks that must be accomplished to attain the overarching objective of air and space dominance. Three functions for the future Air Force emerged from the third and final affinity grouping: awareness, reach, and power. Awareness is specified by the tasks detect, understand, and direct. To have reach requires the ability to deploy, maintain, and replenish. *Power* comes from the ability to engage and survive. The Analysis team adopted the following definitions for these three functions:

> *Awareness* — knowledge, understanding, or cognizance of some thing or situation through alertness in observing, detecting, and identifying, so as to enable, direct, and communicate an informed decision.

Reach — ability to move to expand the range or scope of influence or effect, and to sustain this influence or effect by maintaining and replenishing.

Power — ability to overtly or covertly affect, control, manipulate, deny, exploit, or destroy targets, including forces, people, equipment, and information, and the ability to survive while affecting targets.

These definitions are based on the tasks in the affinity diagrams upon which the functions were built (fig. 2-10), and they suggest that the critical functions of air and space forces in the future do not differ significantly from the functions of today. Where the future begins to diverge from the present is in the detailed means (i.e., tasks and subtasks) by which these functions are accomplished.

Two critical implications appear when the requirement for a set of functions in a value model are mutually exclusive and collectively exhaustive. First, these three **Air Force 2025** functions should encompass every future air and space force operational activity. Second, awareness, reach, and power are the only operational activities that contribute to the overarching objective of air and space dominance.

Once the functions were developed, the bottom-up evolution of the subtasks, tasks, and functions in the **Foundations 2025** value model was complete. Figure 2-13 depicts the entire framework of mutually exclusive and collectively exhaustive functions, tasks, and subtasks to be accomplished by future air and space forces. Next, force qualities, measures of merit, and scoring functions had to be added to the framework to finalize the model.

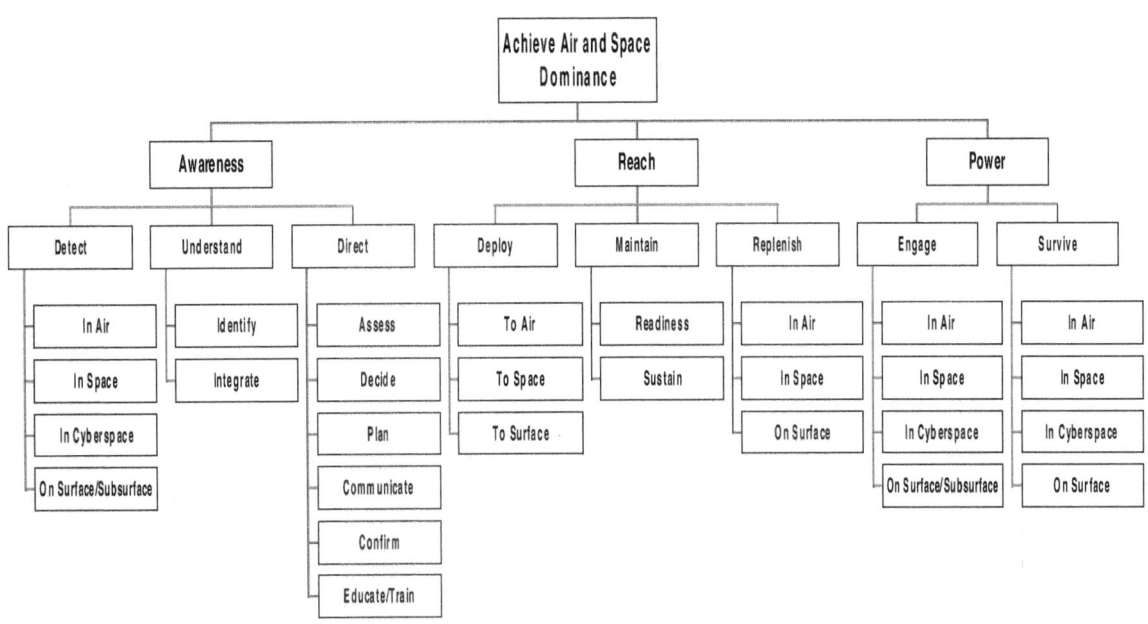

Figure 2-13. *Foundations 2025* **Value Model**

Force Qualities

Though the framework shown in figure 2-13 represented a major breakthrough, it was not a complete value model. The next step for the Analysis team was to meet with each of the *Air Force 2025* white paper writing teams for a second time to determine force qualities, based on the teams' operational expertise, research, and thoughts about the future. Force qualities are generally adjectives, since they characterize a system's ability to accomplish a task or subtask. In many cases, the desired force qualities of a future force did not differ from qualities expected of today's force. For example, the force qualities associated with the subtask *identify* were *accurate, timely,* and *traceable*. The goal was to identify only the most important force qualities for each subtask.

Some common themes emerged in the force quality development process. For almost every subtask, *timely* was a desired force quality. *Unobtrusive* was another commonly occurring force quality; it is listed under every *detect* and *engage* subtask. Clearly, the ability to watch without being watched, and act without being observed, are desirable qualities of future air and space forces. Lastly, most of the writing teams felt that flexibility was important across the spectrum of tasks. Several force qualities address flexibility: multirole, reusable platform, desired lethality, and sensor variety.

These force qualities and measures of merit were continually refined during a succession of meetings. After working with each *Air Force 2025* white paper writing team, the Analysis team was reduced the list of force qualities from the initial number of about 1,200 to the final number of 134. There are about five force qualities for each subtask. The largest number of subtask force qualities was nine, the fewest was two. Figure 2-14 through figure 2-16 show the final force qualities for *Foundations 2025,* organized under the functional categories of *awareness, reach,* and *power*.

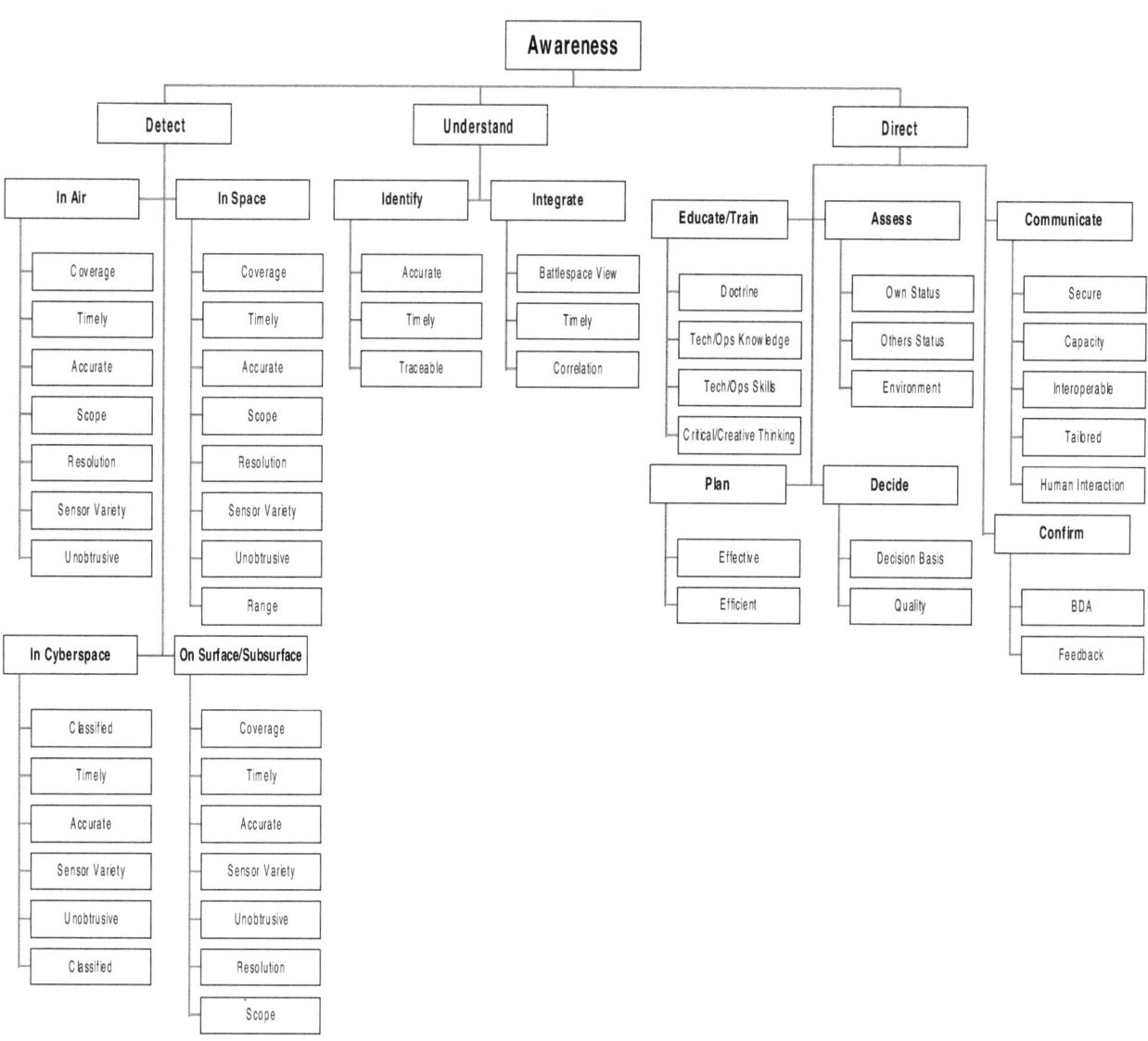

Figure 2-14. *Awareness* **Tasks, Subtasks, and Force Qualities**

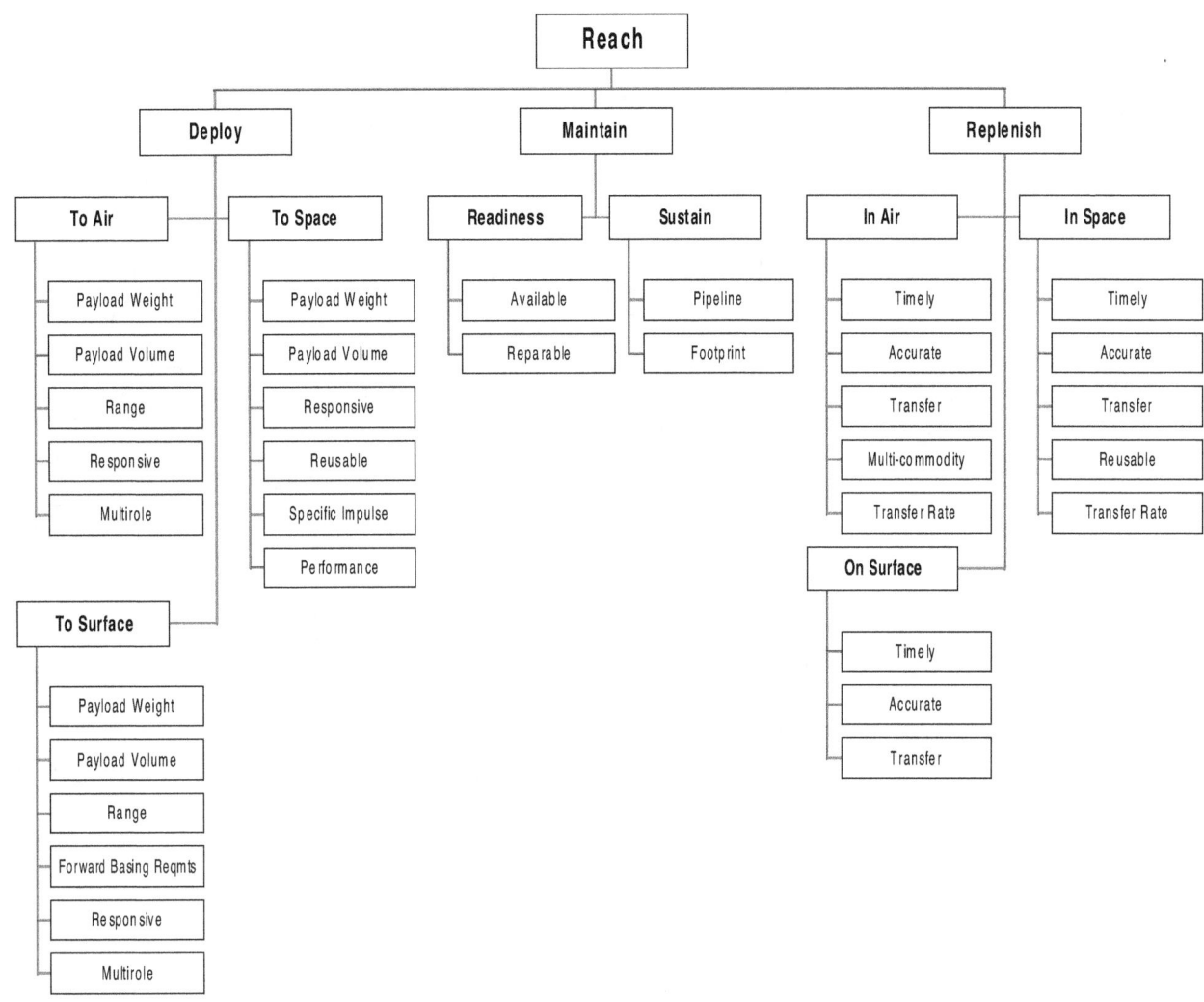

Figure 2-15. *Reach* **Tasks, Subtasks, and Force Qualities**

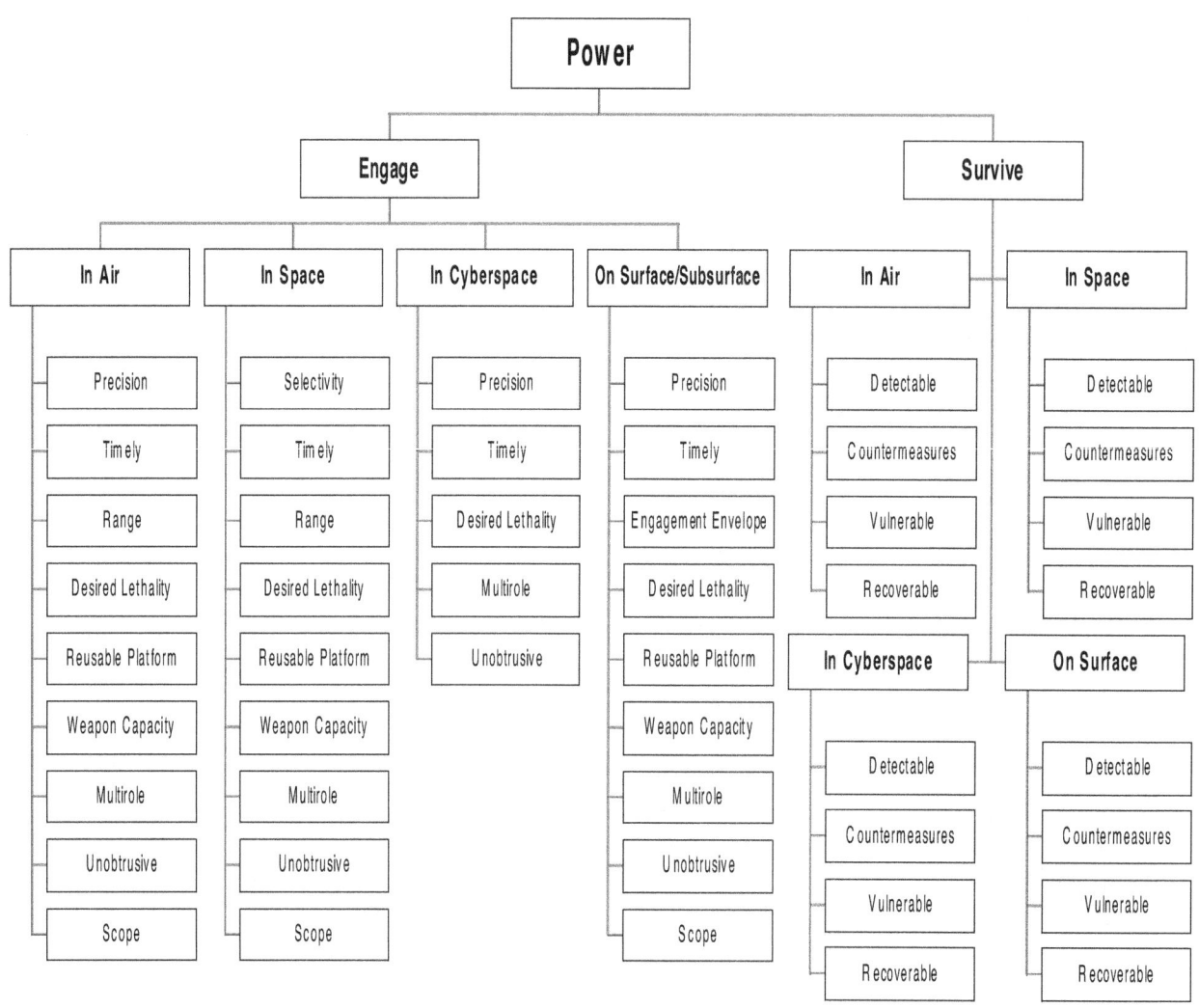

Figure 2-16. *Power* Tasks, Subtasks, and Force Qualities

Measures of Merit and Scoring Functions

Corresponding measures of merit were developed at the same time the Analysis team met with the *Air Force 2025* writing teams to determine force qualities. Each force quality had a measure of merit to calibrate system performance. For example, a force quality of the subtask *deploy to air* was *range*, and the corresponding measure of merit was *miles*. The measures of merit became the domains (horizontal axis) of the scoring functions used to evaluate the capabilities of future systems.

The Analysis team again worked closely with the ***Air Force 2025*** writing teams to build the scoring functions associated with the teams' respective portions of the value model. The result—134 detailed functions that quantify operational values—represent an important analytical accomplishment. They span the spectrum of air and space operations, and, as such, serve as a wealth of information for mission area analysts and weapon system developers. Appendix A contains the complete set of the ***Air Force 2025*** measures of merit and scoring functions.

Foundations 2025 represents five important analytic advances. First, the collection of scoring functions serves as an invaluable resource, even outside the ***Air Force 2025*** study. Second, the use of verbs to specify tasks was a useful step in the value model evolution. Third, the bottom-up approach used in developing ***Foundations 2025*** was significant because no *a priori* assumptions were made and no preconditions were established. Building from the bottom up allowed ***Foundations 2025*** to be free from institutional bias, an outcome necessary to capture the visionary thinking of ***Air Force 2025***. Fourth, ***Foundations 2025*** is a robust value model. With five tiers consisting of an overarching objective, three functions, eight tasks, 29 subtasks, and 134 force qualities (each with a corresponding measure of merit and scoring function)—and all weighted across six alternate futures—the model can be used to evaluate diverse systems. Finally, ***Foundations 2025*** is cast further into the future than any other known military value model.

After the ***Foundations 2025*** development was completed, the next step in the ***Air Force 2025*** OA was to use the model to evaluate systems. The ***Air Force 2025*** white papers provided the key information for identification and definition of the systems.

System Identification

Following a thorough review of the ***Air Force 2025*** white papers, the Analysis team identified 43 unique high-leverage systems. For this operational analysis, a system was defined to be "a functionally related group of elements that performs a mission or task." Although some of the identified systems were extracted from a single white paper, many systems, particularly those involving the collection and management of information, were composites drawn from capabilities detailed in several of the papers. For example, the use of a "personal digital assistant" was a key element of 10 white papers on issues ranging from special operations to general education and training. In several of the papers, such as the one entitled "A Contrarian View of Strategic Aerospace Warfare," no systems could be identified. In these cases, the papers contained a general framework for doing business in given mission areas without a level of detail required for technology identification.

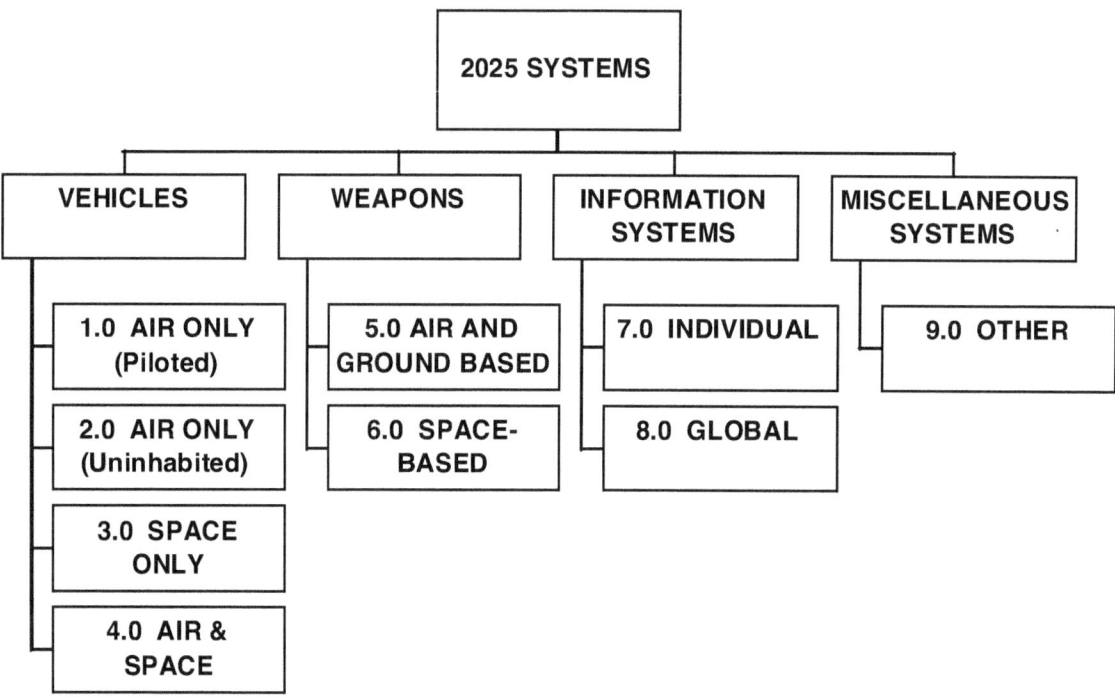

Figure 2-17. System Functional Hierarchy

The 43 systems are listed in table 1, categorized by the major functional areas depicted in figure 2-17. The full descriptions of these systems are found in appendix B.

Table 1

Identified Systems

1.0	Vehicles - Air Only (Piloted)
	1.1 Hypersonic Attack Aircraft
	1.2 Fotofighter
	1.3 Container Aircraft
	1.4 Lighter-than-Air Airlifter
	1.5 Supersonic Airlifter
	1.6 Stealth Airlifter
	1.7 Global Transport Aircraft
2.0	Vehicles - Air Only (Uninhabited)
	2.1 Strike UAV
	2.2 Reconnaissance UAV
	2.3 Uninhabited Combat Air Vehicle
	2.4 Precision Delivery System
	2.5 UAV Mothership
	2.6 Exfiltration Rocket
3.0	Vehicles - Space Only
	3.1 Orbital Maneuvering Vehicle
	3.2 Orbital Combat Vehicle
	3.3 Satellite Bodyguards
4.0	Vehicles - Air and Space
	4.1 Piloted SSTO Transatmospheric Vehicle
	4.2 Uninhabited Air-Launched Transatmospheric Vehicle
5.0	Weapons - Air and Ground-Based
	5.1 Adjustable Yield Munition
	5.2 Advanced Air-to-Air Missile
	5.3 Airborne High-Power Microwave Weapon
	5.4 Standoff Hypersonic Missile
	5.5 Attack Microbots
	5.6 Airborne Holographic Projector
	5.7 Hybrid High-Energy Laser System
6.0	Weapons - Space-Based
	6.1 Global Area Strike System
	6.2 Space-Based Kinetic Energy Weapon
	6.3 Space-Based High-Power Microwave Weapon
	6.4 Space-Based High-Energy Laser
	6.5 Solar-Powered High-Energy Laser System
	6.6 Solar Energy Optical Weapon
	6.7 Asteroid Mitigation System
7.0	Information Systems - Individual
	7.1 Spoken Language Translator
	7.2 Personal Digital Assistant
	7.3 Virtual Interaction Center
8.0	Information Systems - Global
	8.1 Global Information Management System
	8.2 Global Surveillance, Reconnaissance, and Targeting System
	8.3 Sensor Microbots
	8.4 Multiband Laser Sensor System
	8.5 Asteroid Detection System
9.0	Miscellaneous Systems
	9.1 Mobile Asset Repair Station
	9.2 Weather Analysis and Modification System
	9.3 Sanctuary Base

Alternate Futures

The *Air Force 2025* Alternate Futures team generated and then analyzed more than 100 candidate drivers deemed to be forces acting on the future. These drivers were then synthesized and consolidated into the three most important drivers to define a strategic planning space in which alternate futures could be cast (fig. 2-18). Functional definitions for each of these three drivers are provided below:

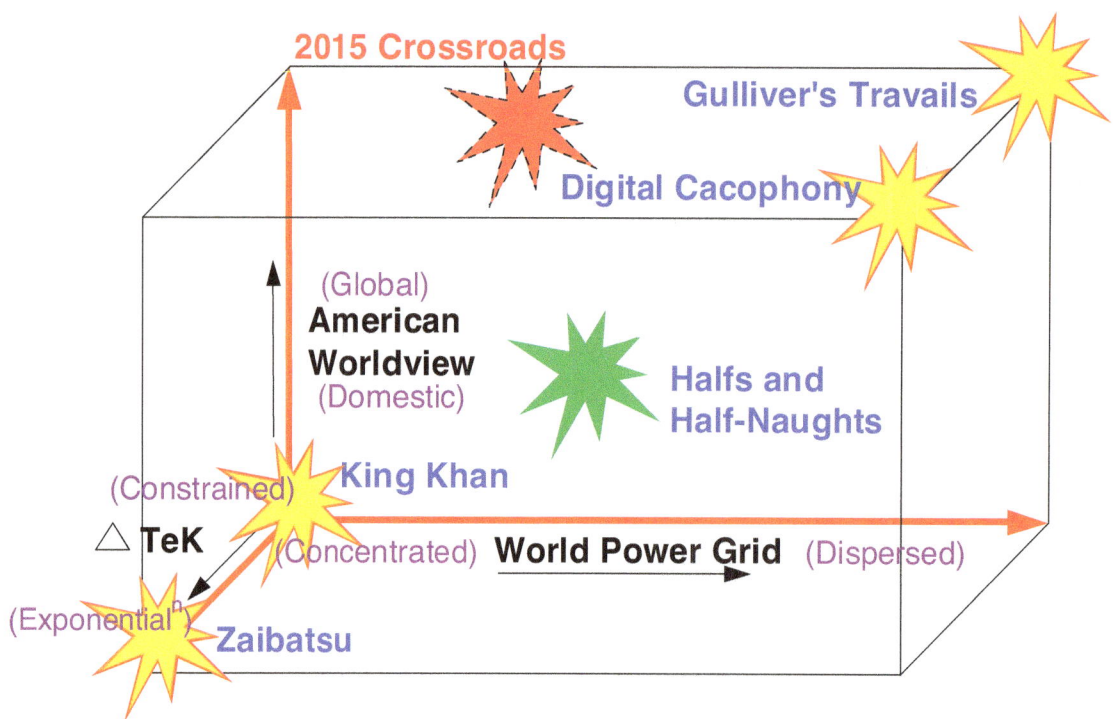

Figure 2-18. Alternate Futures Planning Space

American Worldview. This driver is the US perspective of the world which determines the willingness and capability to interact with the rest of the world. American worldview captures the dominant US focus regarding international affairs. The US can be primarily internally focused, perhaps even isolationist, or the US can be actively engaged in activities around the world. The poles of American worldview are domestic and global.

ΔTeK. This driver is the differential in the rate growth, proliferation, leverage, and vitality of scientific knowledge and technical applications and their consequences. ΔTeK describes the rate of change in both the proliferation and advancement of technology. The two poles of ΔTeK are constrained and exponential. Constrained ΔTeK implies technology is advancing at an evolutionary rate and its availability is limited to a relatively few number of actors. Exponential ΔTeK occurs when there are revolutionary breakthroughs in technology that are rapidly proliferated throughout the world.

World Power Grid. This driver describes the generation, transmission, distribution, and control of power throughout the world. This power is a combination of economic, political, and information sources of power as well as military strength. The two poles of this driver are concentrated and dispersed. A concentrated world power grid exists when few actors have the means or will to influence others. When a myriad of groups or individuals can change the future, the World Power Grid is dispersed.

Six alternate futures were chosen from this planning space to provide a diverse set of future conditions against which to evaluate the proposed air and space systems. Four futures are extremes: Gulliver's Travails, Zaibatsu, Digital Cacophony, and King Khan. The world of Halfs and Half-Naughts was chosen for its centrality. Finally, the *2015 Crossroads* future provides a conservative bridge between today and 2025.

In Gulliver's Travails, the US is overwhelmed with worldwide commitments, counterterrorism and counterproliferation efforts, humanitarian operations, and peacekeeping operations. In Zaibatsu, multinational corporations dominate international affairs, loosely cooperating to create a relatively benign world. Digital Cacophony is the most technologically advanced world resulting in great power and independence for the individual but also creating a world of social isolation, fear, and anxiety. *King Khan* is a world where US dominance has waned due to domestic problems, an economic depression, and overshadowing by a rising Asian colossus. The world of Halfs and Half-Naughts is dominated by conflict between the "haves" and "have-nots" and dynamically changing social structures and security conditions. *2015 Crossroads* uses programmed forces from 1996–2001 to fight a major conflict; it presents the US with a strategic challenge in 2015 that could lead to any of other alternate futures by 2025.

These six alternate futures provided the fulcrum against which the *Air Force 2025* Operational Analysis was applied to determine which of the many systems proposed by the study participants had merit and, hence, should be pursued by the United States Air Force to ensure air and space dominance in the future.

Scoring the Systems

This section describes how **Foundations 2025** was used to evaluate future air and space systems. The process had two steps the first step assign weights to the model hierarchy, and the second computes performance scores using scoring functions. The scoring in this section provides the reader with an example of the score computation process; the results of the actual *Air Force 2025* systems evaluation are described in chapter 3.

Weighting the *Foundations 2025* Value Model across Alternate Futures

The first step in using the 2025 value model is for the decision maker to determine the relative importance of the functions, tasks, subtasks, and force qualities. As described in the previous section, the decision maker weights functions, tasks, subtasks and force qualities. Because different futures dictate a different set of required air and space capabilities, the Analysis team obtained value model weights from the *Air Force 2025* participants for the range of potential future worlds postulated by the *Air Force 2025* Alternate Futures team. For each alternate future, the Analysis team used two sets of weights. The first, denoted "AU Team Weights," is an average of the weights assigned by student members of the *Air Force 2025* white paper writing teams. The second, denoted "Alt Futures Weights," is the weights provided solely by the Alternate Futures team. In general, the Alernatet Futures Weights exhibited greater variation across futures than did the AU Team Weights. Weights were held constant for the force qualities and measures of merit because they were not expected to vary much across possible futures. The weights for each future are contained in appendix C.

Technology Identification and Scoring

Once the 43 unique systems contained in the white papers were identified, the Analysis team qualitatively analyzed each system to identify which technology areas would be key to achieving the stated system capabilities. Only those technology areas needing development were considered. For example, if a specific technology area was critical to a given system's capability, but no new advances were needed in this area for the system to achieve its full capability, then this technology area was not identified as high leverage for this particular system.

The team felt that it was highly desirable to identify and group technologies according to a well-known "gold-standard." Thus, the DOD document, *The Militarily Critical Technologies List* (MCTL), was used as the basis for key technology identification in each system. For the 43 systems evaluated, a total of 43 key technology areas were identified. Full descriptions of each technology area, as paraphrased from the MCTL, can be found in appendix D.

To eventually rank technologies by their impact on future air and space capabilities, the team assigned a relative weight to each technology embedded in a particular system. The weights selected sum to 100 for each system and so can be thought of as percentages of the system's dependence on each technology needing development. For example, the five piloted single-stage-to-orbit (SSTO) transatmospheric vehicle (TAV) technologies were weighted as follows:

Technology Area	Weight
Aerospace Structures and Systems	25
High-Energy Propellants	25
RAMjet, SCRAMjet, Combined Cycle Engines	20
Advanced Materials	20
High-Performance Computing	10

In this case, since the primary mission of the piloted SSTO TAV is to travel between the surface and low-earth orbit, the highest leverage technology areas were those of the vehicle's primary propulsion and structural subsystems. Each of these areas were evaluated to be essentially equal in importance. The fifth technology area, high performance computing, was added not necessarily because of vehicle requirements, but because the design process for this type of vehicle will take some advances in computing power. Without advances in high-performance computing, the design process for a TAV with this capability would be impaired. This methodology makes it possible to score each of the systems.

Once the system-versus-technology matrix is developed, the procedure for scoring the technologies is straightforward. The contribution of technology to each system is multiplied by the system value, and the resulting products are summed across all systems. The result is a set of technology scores (normalize to a maximum score of 100) that takes into account both the technologies' degree of contribution to future air and space systems and the importance of those systems to air and space operations. This scoring was then repeated for each alternate future since the system values changed with each future.

After each technology area had been identified, one additional question remained, "Who, the DOD or the commercial sector, is the key player in developing each of the 43 technology areas?" AFIT's Graduate

School of Engineering assembled a committee from its senior staff to consider this question. This committee qualitatively evaluated each technology area to determined the appropriate key developer. They further ascertained the direction of each developmental effort, whether from the DOD to the commercial sector, from the commercial sector to the DOD, or remaining constant.

Notes

[1] Ralph L. Keeney, *Value-Focused Thinking: A Path to Creative Decisionmaking* (Cambridge, Mass.: Harvard University Press, 1992).
[2] **SPACECAST 2020** *Operational Analysis* (Maxwell AFB, Ala: Air University Press, 1994).
[3] Keeney, 3.
[4] Ibid., 6.
[5] Ibid., 7.
[6] Ibid., 33.
[7] Ibid., 112.
[8] R. T. Clemen, *Making Hard Decisions: An Introduction to Decision Analysis* (Boston, Mass.: PWS-Kent, 1991).
[9] Keeney.
[10] Wayne Winston, *Operations Research: Applications and Algorithms,* 3d ed. (Belmont, Calif.: Duxbury Press, 1993).
[11] Keeney, 78.
[12] Ibid.
[13] Clemen, 435.
[14] Ibid., 435–436.
[15] Keeney, 129-154.
[16] Clemen, 435–36.
[17] *A National Security Strategy of Engagement and Enlargement* (Washington, D. C.: The White House, February 1995).
[18] *National Military Strategy of the United States of America, 1995* (Washington, D. C.: US Government Printing Office).
[19] *Defense Planning Guidance* (Washington, D. C.: Office of the Secretary of Defense, 1995).
[20] *Global Reach, Global Power: A White Paper*, Department of the Air Force, June 1990.
[21] *Global Presence*, Department of the Air Force, 1995.
[22] Glenn A. Kent and William E. Simons, *A Framework for Enhancing Operational Capabilities*, R-4034-AF (Santa Monica, Calif.: RAND 1991).
[23] *Air Force Doctrine Document: Air Force Basic Doctrine*, 15 August 1995, first draft (Washington, D. C.: US Air Force, 1995).
[24] *Joint Vision 2010, America's Military: Shaping the Future* (Washington, D. C.: Joint Chiefs of Staff, January 1996).
[25] *Cornerstones of Information Warfare* (Washington, D. C.: US Air Force, 1995).

[26] Michael Brassard and Diane Ritter, *The Memory Jogger II: A Pocket Guide of Tools for Continuous Improvement and Effective Planning* (Methuen, Mass.: Growth Opportunity Alliance of Lawerance/QPC, 1994), 13–14.

[27] Cotton, Bob, *The Cyberspace Lexicon: An Illustrated Dictionary of Terms From Multimedia to Virtual Reality* (London: Phaedon, 1994).

Chapter 3

Results

A team of technical and operational experts, the Analysis team, scored all 43 systems against each metric in ***Foundations 2025***. The team followed a consensus-seeking approach to obtain each score. The team was not permitted to know the shape of the scoring function and was tasked to place a score for each metric along the horizontal axis of each function shown in figures A-5 through A-32 for each system. The analysis was performed using Logical Decisions for Windows© software running on a Pentium-based PC.

The results of the system scoring are summarized in figure 3-2 through figure 3-6. The vertical axis is the score from the system evaluation on a scale of 0 to 100, where a system score of zero equates to no score on any of the 134 scoring functions. The horizontal axis is a rank ordering of the systems according to the Analysis team's assessment of the relative amount of technical challenge to develop each system. The system scores are shown for two separate sets of weights—the AU Team Weights and the Alternate Futures Weights—for all six of the alternate futures. (figure 3-2 and figure 3-3).

Each system's scores for the various futures are plotted and connected with a line to show the variation of that system's score across the alternate futures. The resulting spread of scores for each system can be regarded as similar to error bars in the results of a statistical sampling technique. In other words, a system's score can be said with high confidence to lie within the range of the points shown. The curved dashed line provides a further reference for comparing systems. In the Analysis team's estimation, systems above the line may have sufficient value to offset the technical challenge of producing such a system. Thus, systems to the left of the charts need less value to be attractive options than systems to the right of the chart, because the difficulty of achieving the capability is much less. The exact location of the line is somewhat arbitrary. It was drawn fairly low, terminating on the right at a value of 20, to identify the most promising systems. (Note: A score of 10-15 may be good for some tasks.)

The highest value systems evaluated in this study are the Global Information Management System (GIMS), Sanctuary Base, Global Area Strike System (GLASS), Global Surveillance and Reconnaissance System (GSRT), and Uninhabited Combat Air Vehicle (UCAV). GIMS has the highest score but has high technical challenge, while GSRT performs some of the functions of GIMS but also with less technical challenge. Because of this, GSRT could be considered a "stepping stone" to GIMS. Both GLASS and UCAV score well because of a strong awareness component to complement their power contributions, and UCAV is the most feasible of all the high-value systems in the near term. Sanctuary base has high value but also the highest technical challenge. It may remain infeasible even beyond 2025. Appendix E contains tables of each system's value score for each future and weight set.

Dividing the figures above into quadrants provides a delineation of the systems into four categories based on value and technical challenge. Figure 3-1 shows representative systems for each quadrant.

Figure 3-1. Representative Systems

Figure 3-4 and figure 3-5 provide a closer look at the highest scoring systems. Limiting the systems displayed to the top 25 percent (11 systems) for the AU Team Weights yields figure 3-4. A comparison of figure 3-4 and figure 3-5 reveals little difference between the overall scores and rankings for the AU Team Weights and the Alternate Futures Weights. The variation between futures is greater for the Alternate Futures Weights, reflecting both the more extreme views of a team immersed in the construction of the alternate futures and the centering effect of averaging the weights of the AU team members. However, the top 11 systems for the AU Team Weights are contained in the top 14 systems for the Alternate Futures Weights, with the Hypersonic Attack Aircraft, Fotofighter, and Sensor Microbots rounding out the 14 systems.

Figure 3-6 shows the relative contribution of the awareness, reach, and power sub-scores to the top 11 systems' scores using the AU Team Weights for the "Halfs and Half Naughts" alternate futures.

It is interesting to note the relationship between the awareness, reach, and power contributions to a system's score and the score variation between alternate futures. Systems that score similarly in awareness, reach, and power (e.g., GLASS in figure 3-6) tend to have the least variation; that is, the line connecting their scores for each future is short. This is because the weighted average of awareness, reach, and power (the overall score) is insensitive to changes in the weights when the awareness, reach, and power scores are of the same magnitude.

The scoring results highlight that a complex system (a system of systems) outscores any of its components. This is because of the additive nature of the scoring functions. The complex system scores more broadly since it contains the capabilities of its components, so it exceeds the score of any single component. Conversely, since component systems are unlikely to score in mutually exclusive areas of the value model, the complex system generally will score less than the simple sum of the component system scores.

Finally, figure 3-7 through figure 3-9 contain graphs similar to those in figure 3-2 and figure 3-3, but for the awareness function, the deploy task of reach, and the power function, respectively, using the AU Team Weights. These figures allow the reader to note the systems which score well for a particular function. For example, figure 3-7 highlights the best systems in terms of the awareness function. Such a level of detail may prove useful when conducting mission area analysis to determine required improvements for specific functional areas. In fact, the software used in this analysis can display the system scores at any level of the value model.

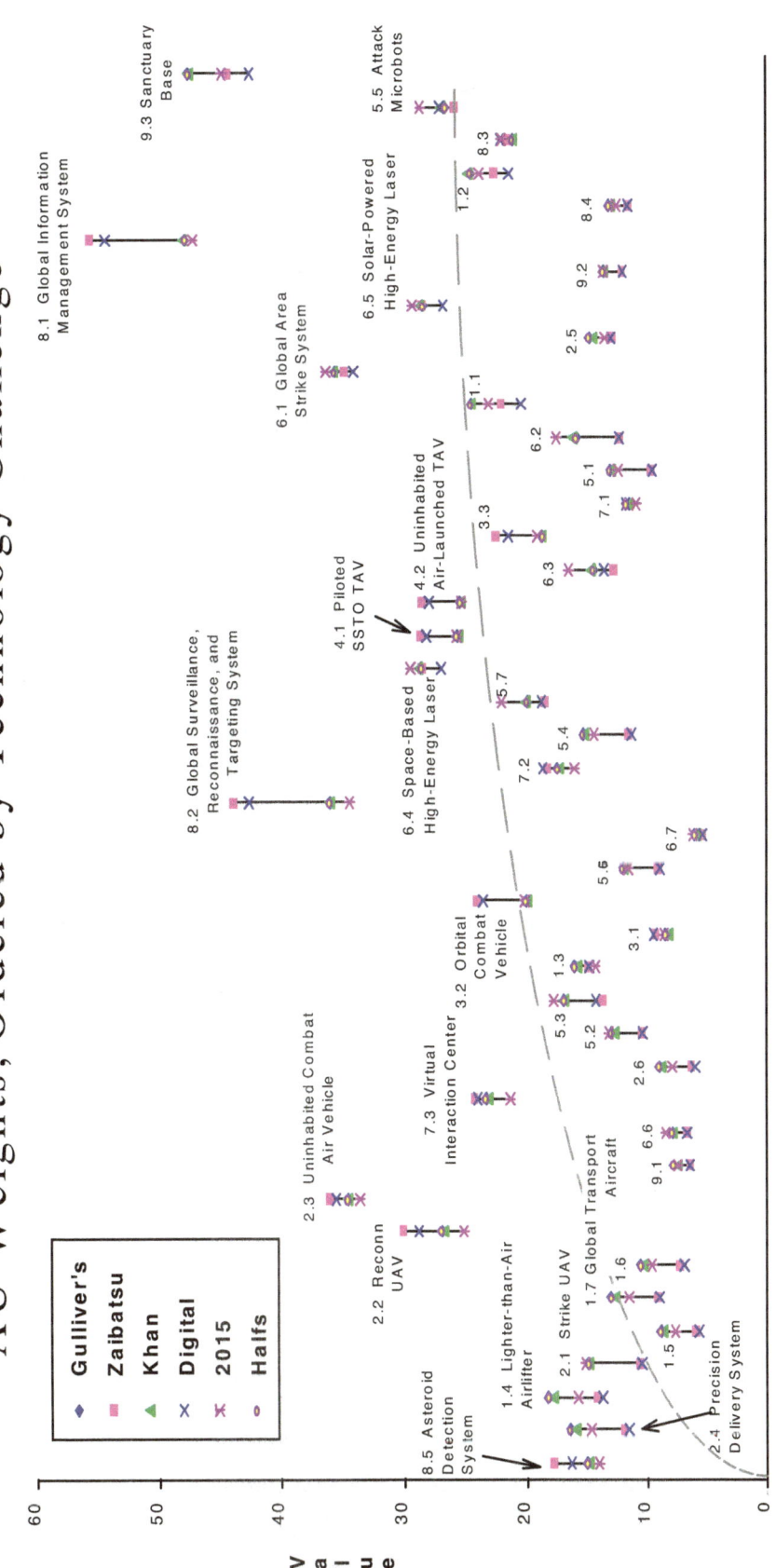

Figure 3-2. Final System Scores - AU Team Weights

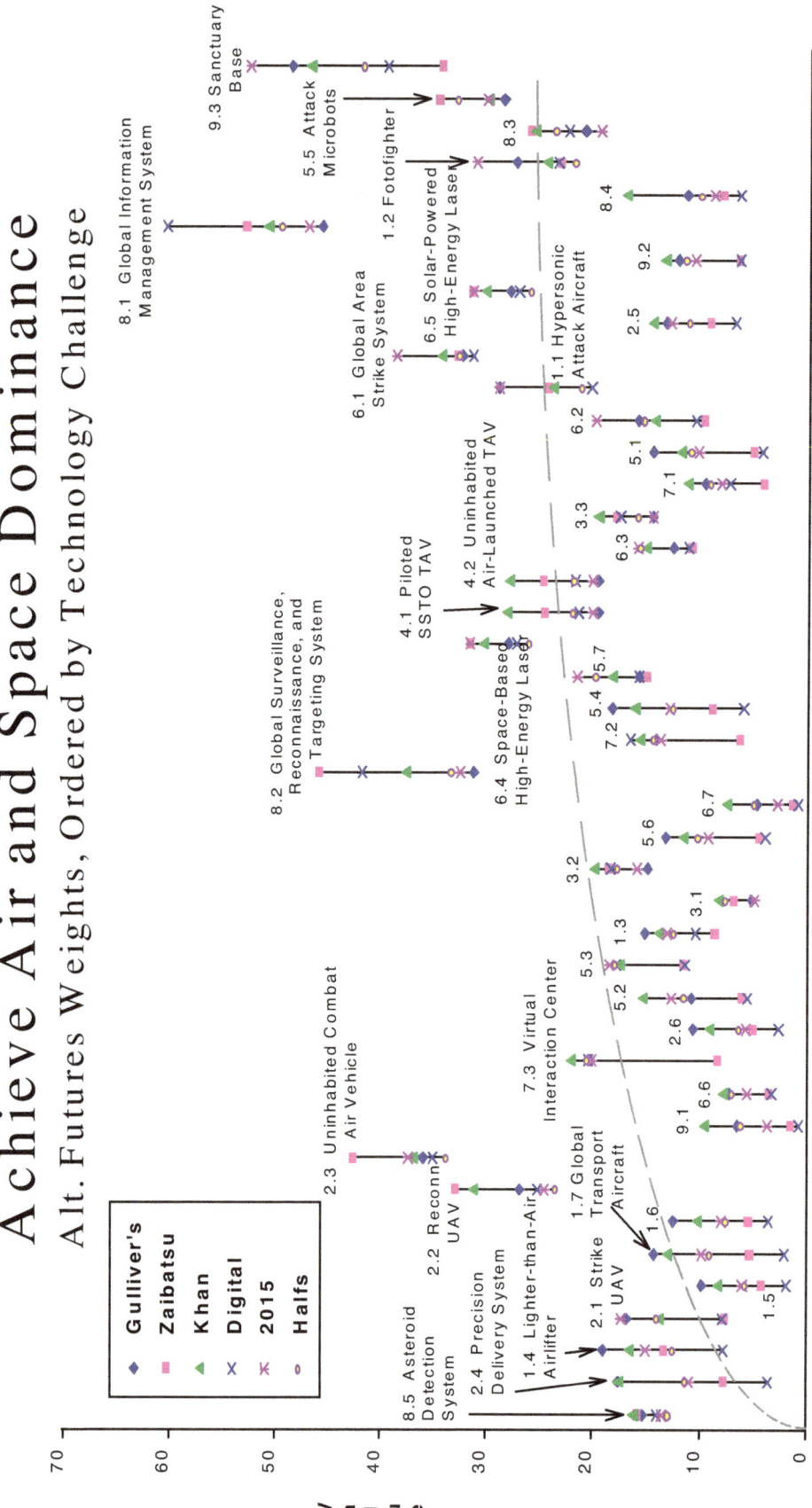

Figure 3-3. Final System Scores - Alt. Futures Weights

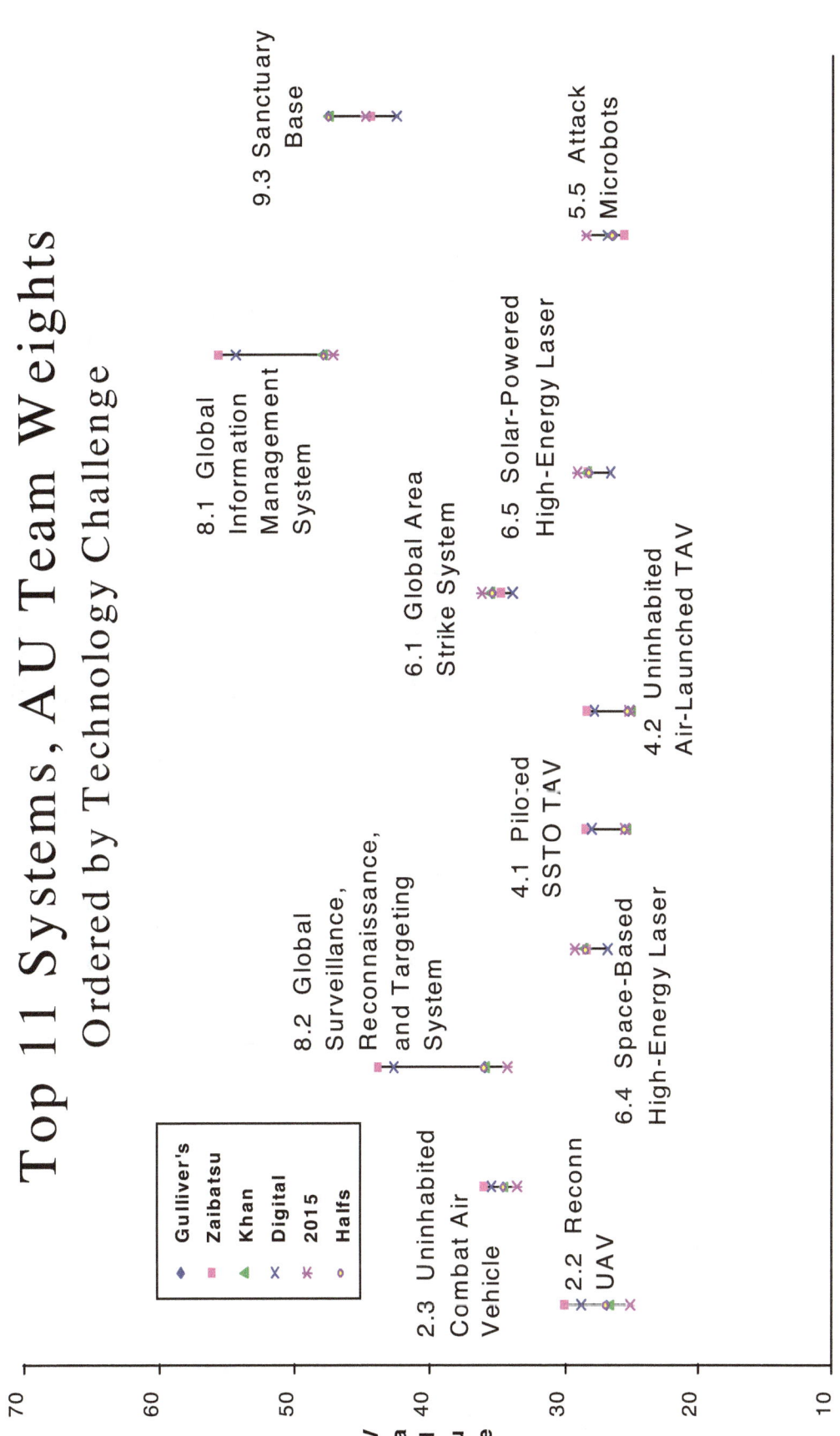

Figure 3-4. Top 11 Systems - AU Team Weights

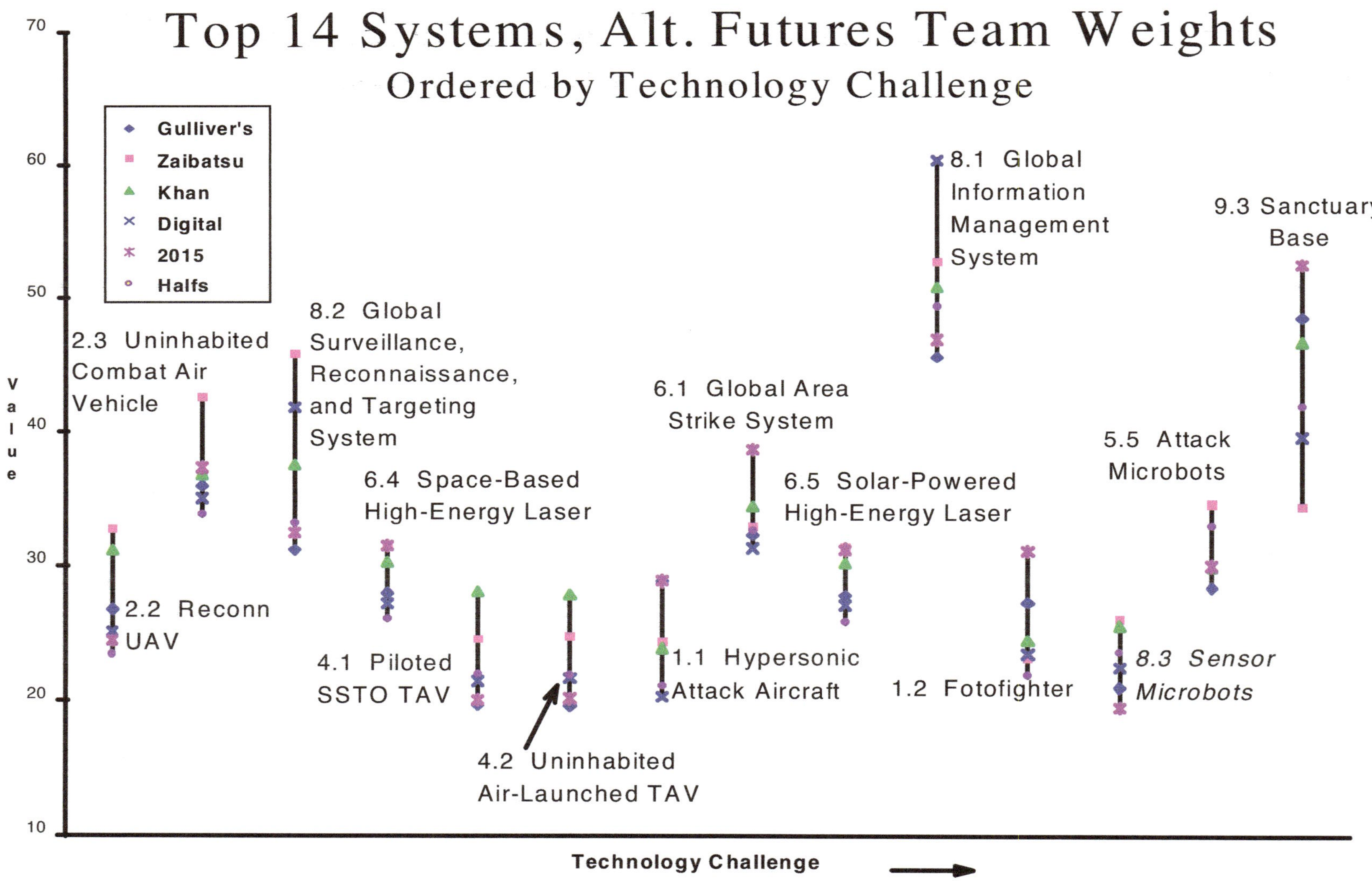

Figure 3-5. Top 14 Systems - Alt. Futures Weights

Top 11 System Rankings
AU Team Weights, "Halfs" Alt Future

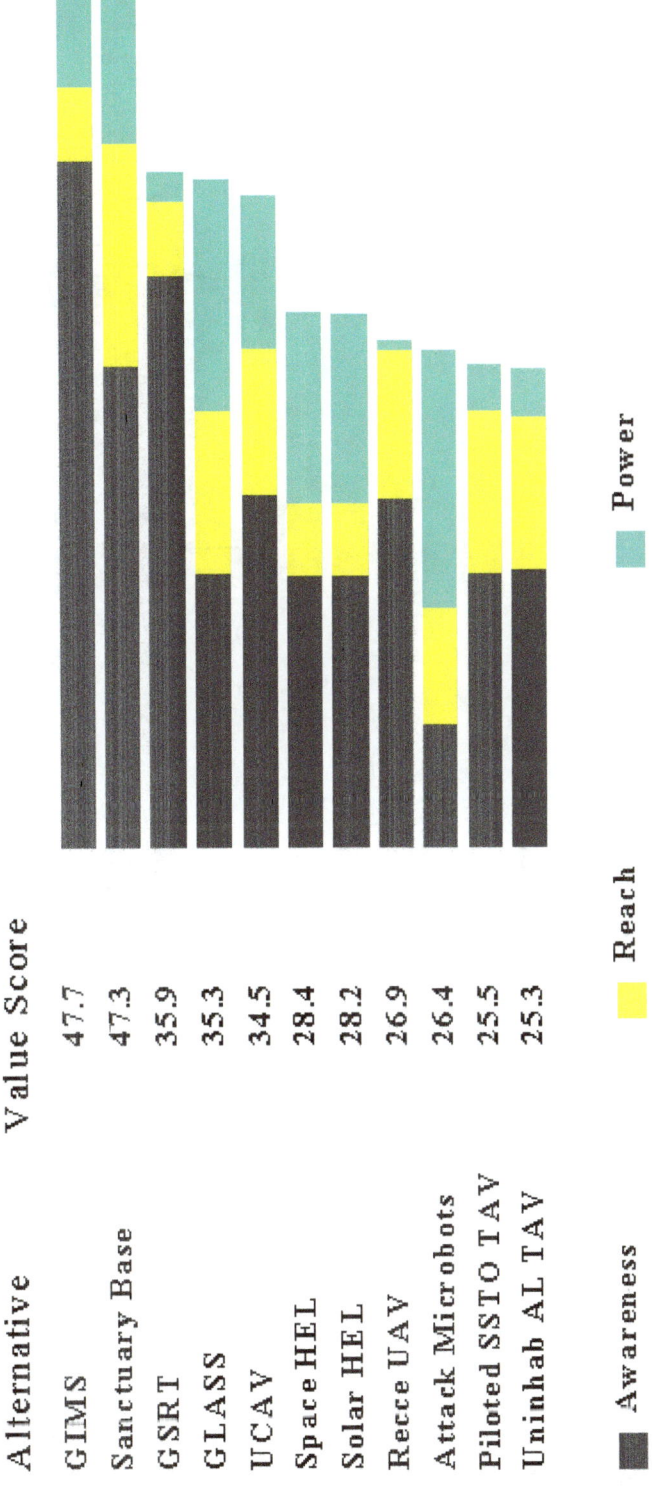

Figure 3-6. Top 11 System Rankings

Figure 3-7. *Awareness* Scores - AU Team Weights

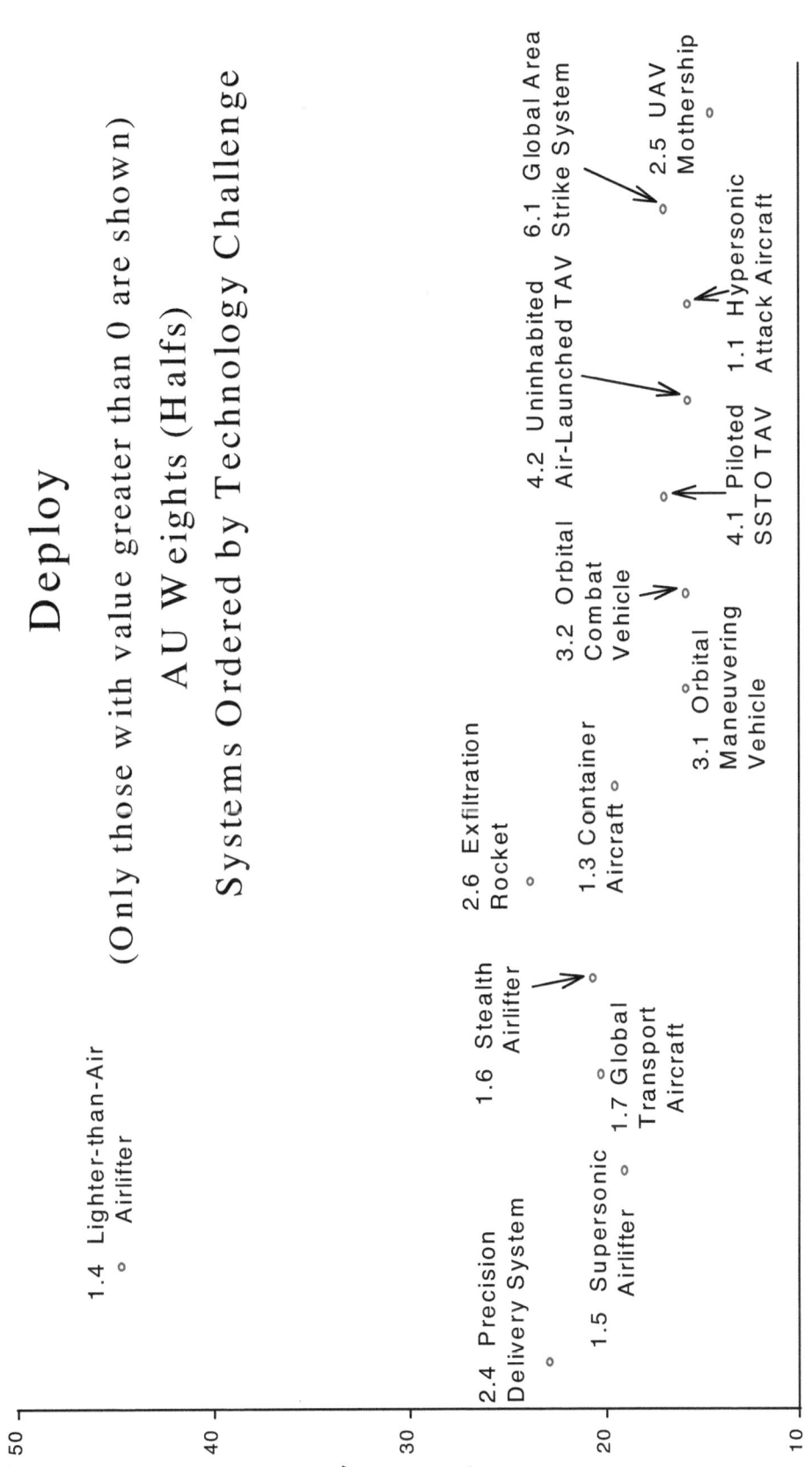

Figure 3-8. Deploy Scores - AU Team Weights, Halfs Future

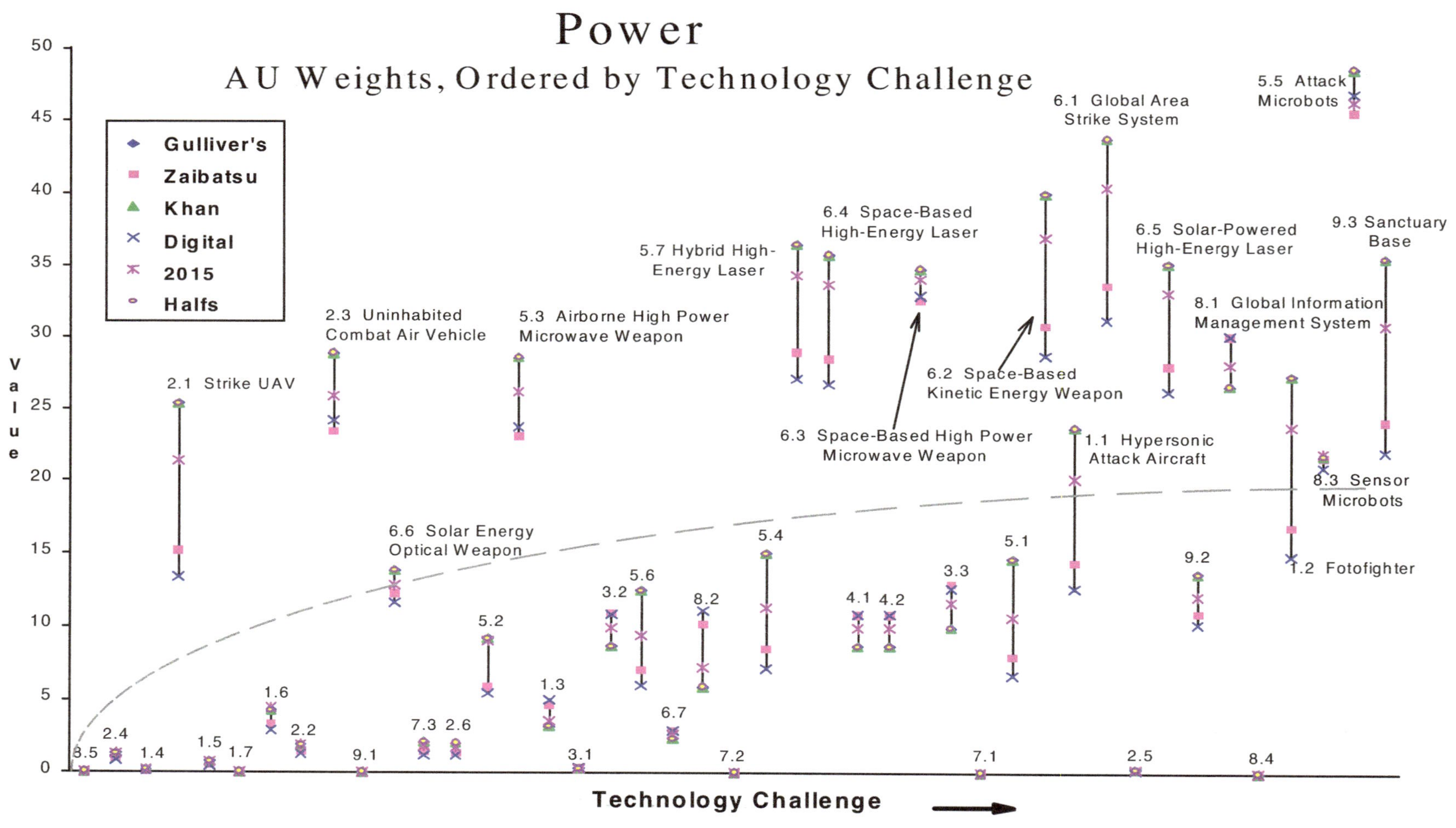

Power
AU Weights, Ordered by Technology Challenge

Figure 3-9. *Power* Scores - AU Team Weights

Scoring the Technologies

The baseline technology assessment is summarized in figure 3-10 for each of the alternate futures. This assessment considers all 43 systems scored and the Value Model was weighted by all AU white paper writing teams. The score for each technology area was calculated by multiplying the percentage dependence of each of the systems on that development technology by the score that system received in the Value Model. The scores were then summed across the all systems with the final result being normalized to a maximum score of 100. These scores are measures of the potential of each enabling technology to improve operational effectiveness in air and space.

Using the "Halfs and Half-Naughts" alternate future, which is placed in the center of the strategic planning space for this study, the technology areas clearly divide themselves into three groupings: the top seven technologies (high leverage), the next fivehnologies (moderate leverage), and the bottom 31 technologies (lesser leverage). Figure 3-11 shows an expanded view of the top two technology groupings for each alternate future.

As a verification of these results, the Analysis team decided to examine the analysis of the technologics by considering their interaction with only the 11 top-scoring systems. These results are shown in figure 3-12 and figure 3-13. Figure 3-12 shows that the three technology groups generally remained, although the top two groupings contain six technology areas each rather than seven and five, respectively, as in the previous case. The six high-leverage technologies all appeared in the previous high-leverage grouping. Further, 11 of the top 12 technologies remained the same. Lastly, seven of the total 43 technology areas were not applicable when the systems considered were narrowed to the 11 top-scoring ones.

Within technology groupings, the rank changed when going from considering all 43 systems to only considering the 11 top-scoring systems. However, with only three exceptions, technology areas did not change their respective groupings. These exceptions were Aerospace Structures (9.5.4) and Vehicle Flight Control (7.3), which both dropped to a lower technology grouping, from high and moderate leverage to moderate and less leverage, and Communications (5.1), which jumped to a higher technology grouping from lesser leverage to moderate leverage. The results of these two assessments are summarized in table 2 for the high and moderate leverage technologies. The numbers in parentheses indicate the appropriate MCTL category which further defines the technology area.

Table 2

Technology Assessment

	ALL 43 SYSTEMS	TOP 11 SYSTEMS
HIGH LEVERAGE TECHNOLOGIES	Power Systems (10.3) Advanced Materials (1.0) Aerospace Structures (9.5.4) High Performance Computing (4.1.1) Micromechanical Devices (2.6) High Energy Propellants (12.7) Data Fusion (4.2.5)	Data Fusion (4.2.5) Power Systems (10.3) Micromechanical Devices (2.6) Advanced Materials (1.0) High Energy Propellants (12.7) High Performance Computing (4.1.1)
MODERATE LEVERAGE TECHNOLOGIES	Artificial Intelligence (4.2.9) High Energy Laser Systems (11.1) Vehicle Flight Control (7.3) Image Processing (4.1.4) Optics (10.2)	High Energy Laser Systems (11.1) Artificial Intelligence (4.2.9) Optics (10.2) Image Processing (4.1.4) Aerospace Structures (9.5.4) Communications (5.1)

A sensitivity study also was performed using the Value Model as weighted by the alternate futures team only. As noted in the previous chapter, these weights had a larger variance than the weights from all AU student teams. The results when all 43 systems were considered are given in figure 3-14 and figure 3-15; figure 3-16 and figure 3-17 show the results when the systems considered were narrowed to the 11 top-scoring ones. Although the variance bars increased in length, the basic conclusions did not change with this new set of weights. Appendix F lists numerical data for both the raw and weighted scores for each technology area.

A common trend among the higher leverage technologies was that they had wide applicability over the systems considered. When all 43 systems were considered, the high leverage technologies scored in at least 13 different systems; the maximum number of systems where any technology area scored was 27. Moderate leverage technologies scored in 8 to 12 different systems. When the systems considered were reduced to the 11 top-scoring ones, the high leverage technologies scored in at least five systems; the maximum number of systems where any technology area scored was nine. Moderate leverage technologies scored in either three or four different systems. In both assessments, High-Performance Computing (4.1.1) was the technology area with the broadest coverage over the systems considered.

After each technology area had been scored, AFIT's Graduate School of Engineering assembled a committee from its senior staff to determined the key technology driver, the DOD or the commercial sector, for that particular area. They further ascertained the direction of each developmental effort, whether from the DOD to the commercial sector, from the commercial sector to the DOD, or remaining constant. Table 3 summarizes the key technology development leaders; two technology areas, 6.9 Other Sensor and 9.8 Other Propulsion, were not considered in this assessment because of the extremely broad coverage of their respective areas.

Table 3

Technology Development Leaders

KEY TECHNOLOGY	DOD LEAD	BOTH DOD & COMM	COMM LEAD
1.0 Materials Technology		X	
2.0 Industrial Production Technology			
2.1 Auto of Ind Process, Sys, and Fact			X--->
2.2 Metal Working and Ind Production			
2.2.1 Num Contr Machine Tools			X--->
2.2.5 Robots, Contr, & Effectors		X	
2.6 Micromechanical Devices		X--->	
4.0 Computer Technology			
4.1 Digital Processing			
4.1.1 High Performance Computing			X
4.1.2 Dyn Training and Simulation		X	
4.1.3 Signal Processing		X	
4.1.4 Image Processing		X	
4.1.6 Speech Processing		X	
4.2 Software			
4.2.4 Hard Real Time Systems	X--->		
4.2.5 Data Fusion	X--->		
4.2.9 Artificial Intelligence			<---X
4.3 Hybrid Computing		X	
5.0 Telecommunications Technology			
5.1 Transmission			X
5.3 Comm Net Mgmt & Control			X
5.4 C3I Systems	<---X		
5.5 Information Security		X	
6.0 Sensors and Elect Combat Technology			
6.1 Acoustic Systems	X--->		
6.2 Optical Sensors	X--->		
6.4 Electronic Combat	<---X		
6.6 Magnetometers	X--->		
6.7 Gravity Meters	X		
6.8 Radar	X--->		
6.9 Other Sensor			
7.0 Nav, Guid, and Vehicle Contr Tech			
7.3 Vehicle and Flight Control		X	
9.0 Propulsion and Vehicular Sys Tech			
9.1 Gas Turbine Propulsion Systems		X	
9.2 RAMjet, SCRAMjet, and CC Engine	<---X		
9.4 Rockets	X (no NASA)	X (if NASA)	
9.5 Aerospace Structures and Systems			
9.5.1 Spacecraft Structures		X	
9.5.2 Non-Chem, High-Isp Prop	<---X		
9.5.4 Aircraft High-Perf Structures	<---X		
9.8 Other Propulsion			

Table 3 (Cont.)

KEY TECHNOLOGY	DOD LEAD	BOTH DOD & COMM	COMM LEAD
10.0 Laser, Optics and Power Sys Tech			
10.1 Lasers		X	
10.2 Optics			X
10.3 Power Systems	X		
11.0 Directed Energy & Kinetic Energy Sys			
11.1 High Energy Laser Systems	<---X		
11.2 High Power Radio Freq Sys	<---X		
11.4 Kinetic Energy Systems			
11.4.2 Kinetic Energy Projectiles	<---X		
11.4.4 KE Platform Management	<---X		
12.0 Munitions Dev & Energetic Mat'l Tech			
12.1 Warheads, Ammo, & Payloads	<---X		
12.7 Mil Explosives (Energetic Mat'l)	X		
13.0 Chemical & Biological Systems Tech			
13.3 CBW Defensive Systems	X		

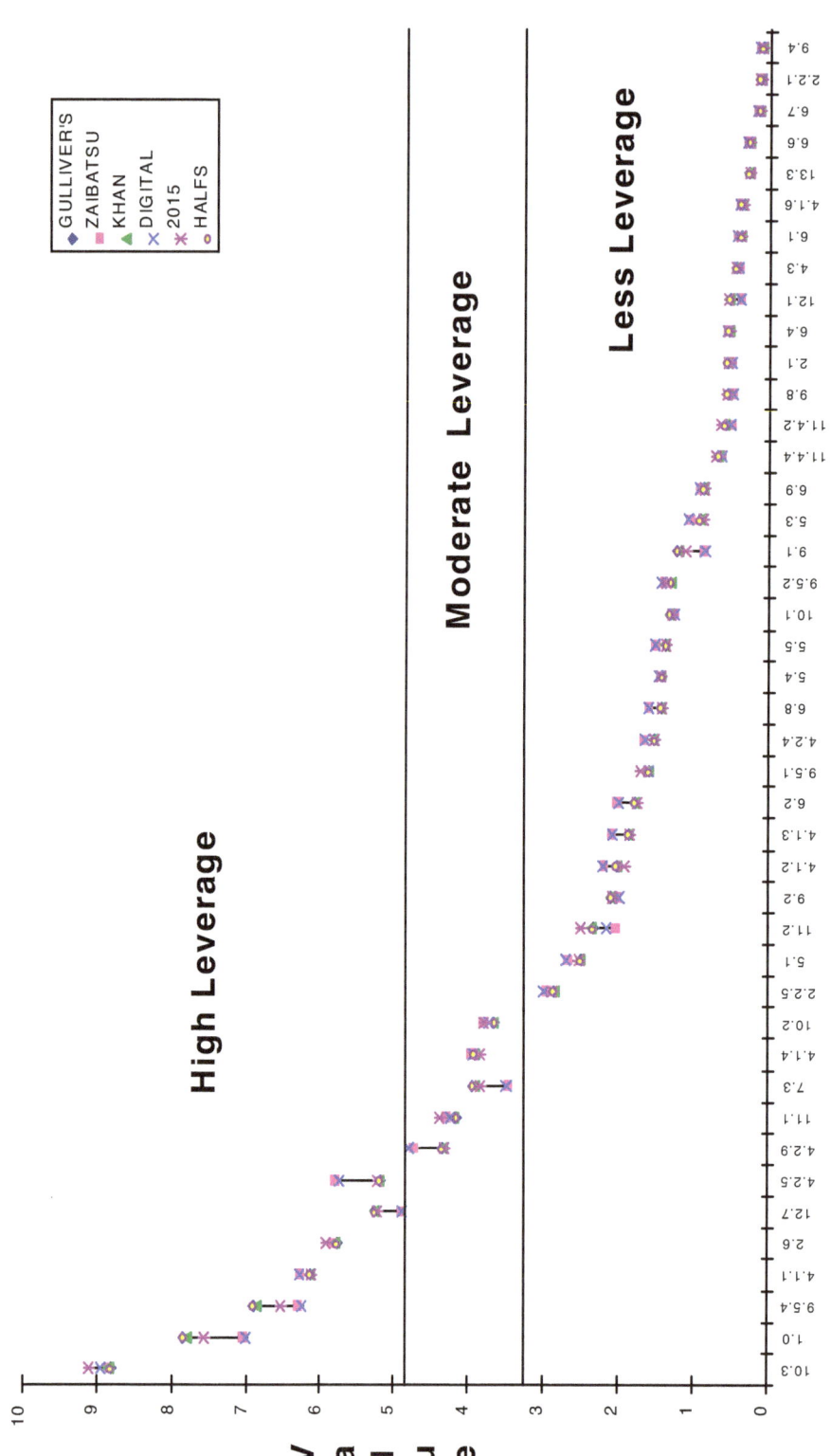

Figure 3-10. Technology Rankings (All 43 Sys, AU Students Wts)

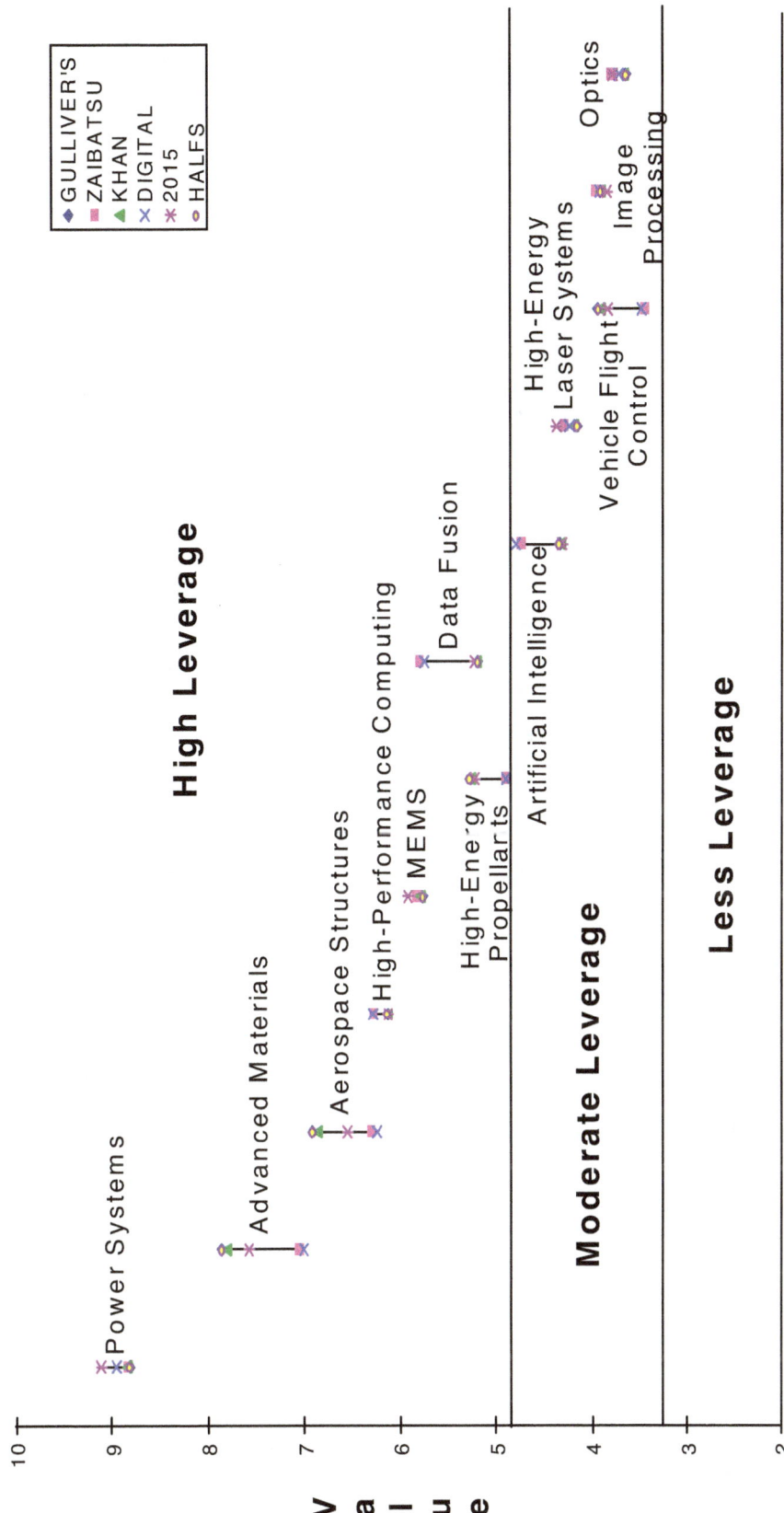

Figure 3-11. Top 12 Technology Rankings (All 43 Sys, AU Students Wts)

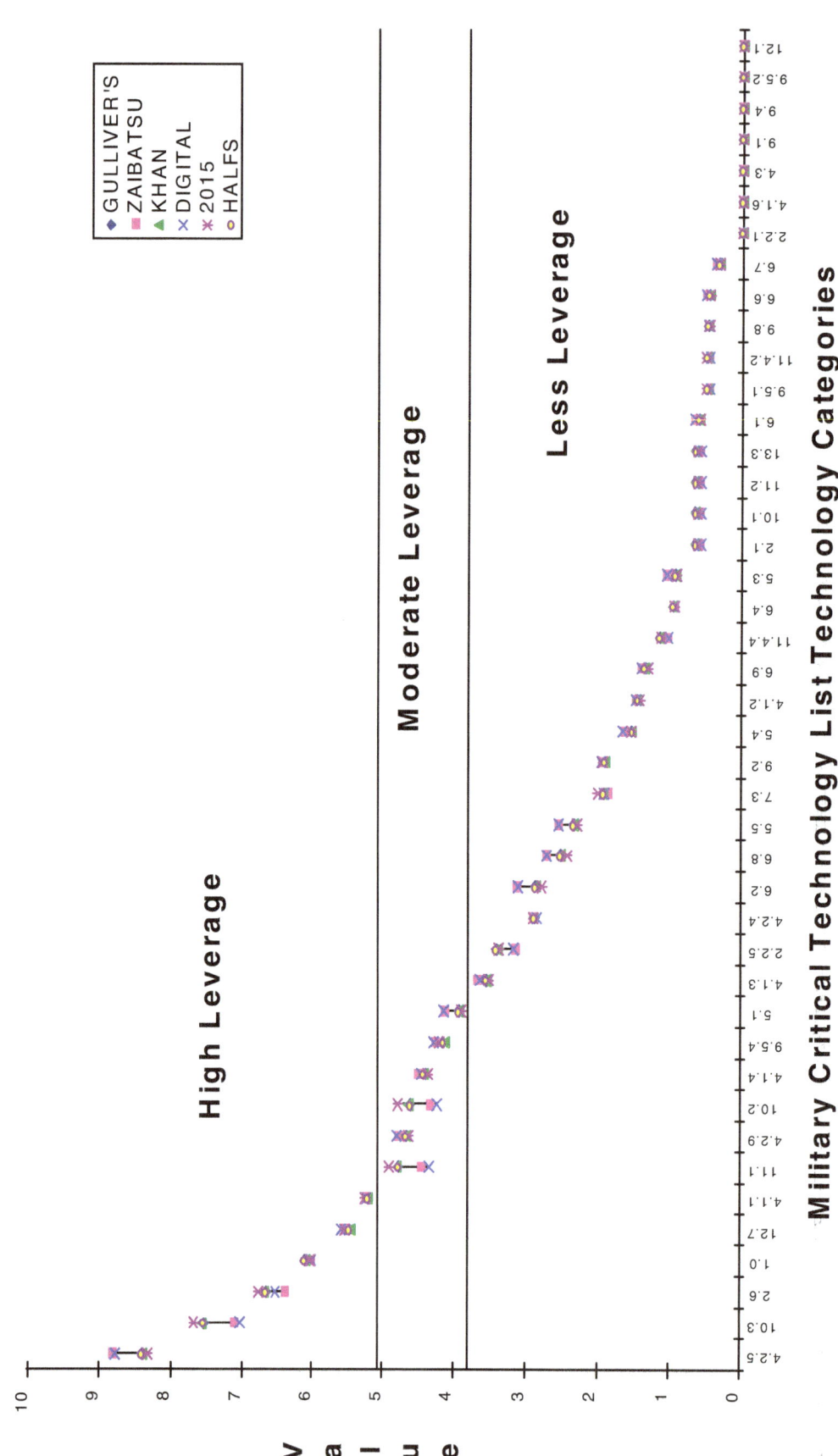

Figure 3-12. Technology Rankings (Top 11 Sys, AU Students Wts)

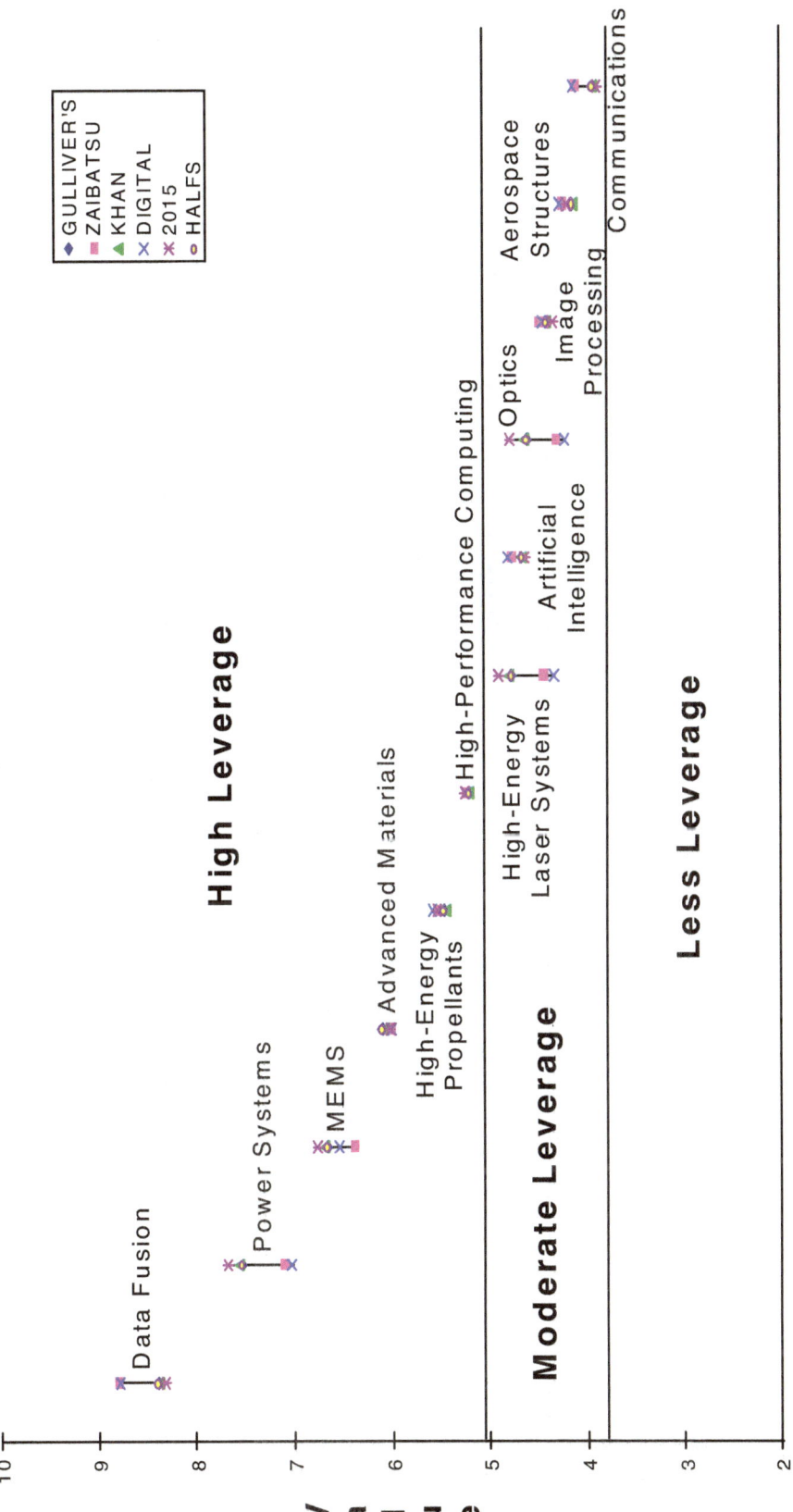

Figure 3-13. Top 12 Technology Rankings (Top 11 Sys, AU Students Wts)

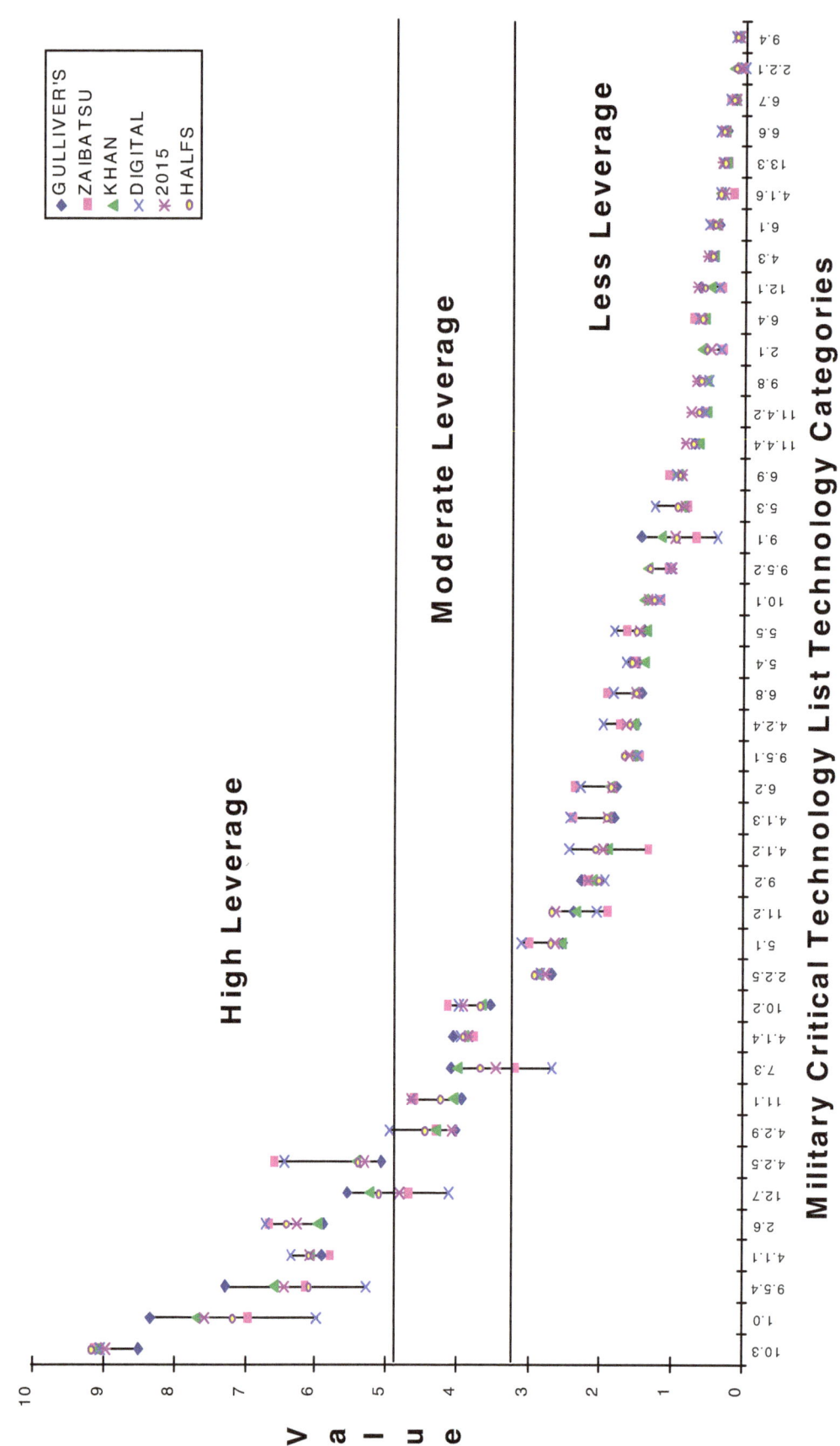

Figure 3-14. Technology Rankings (All 43 Sys, Alt Fut Team Wts)

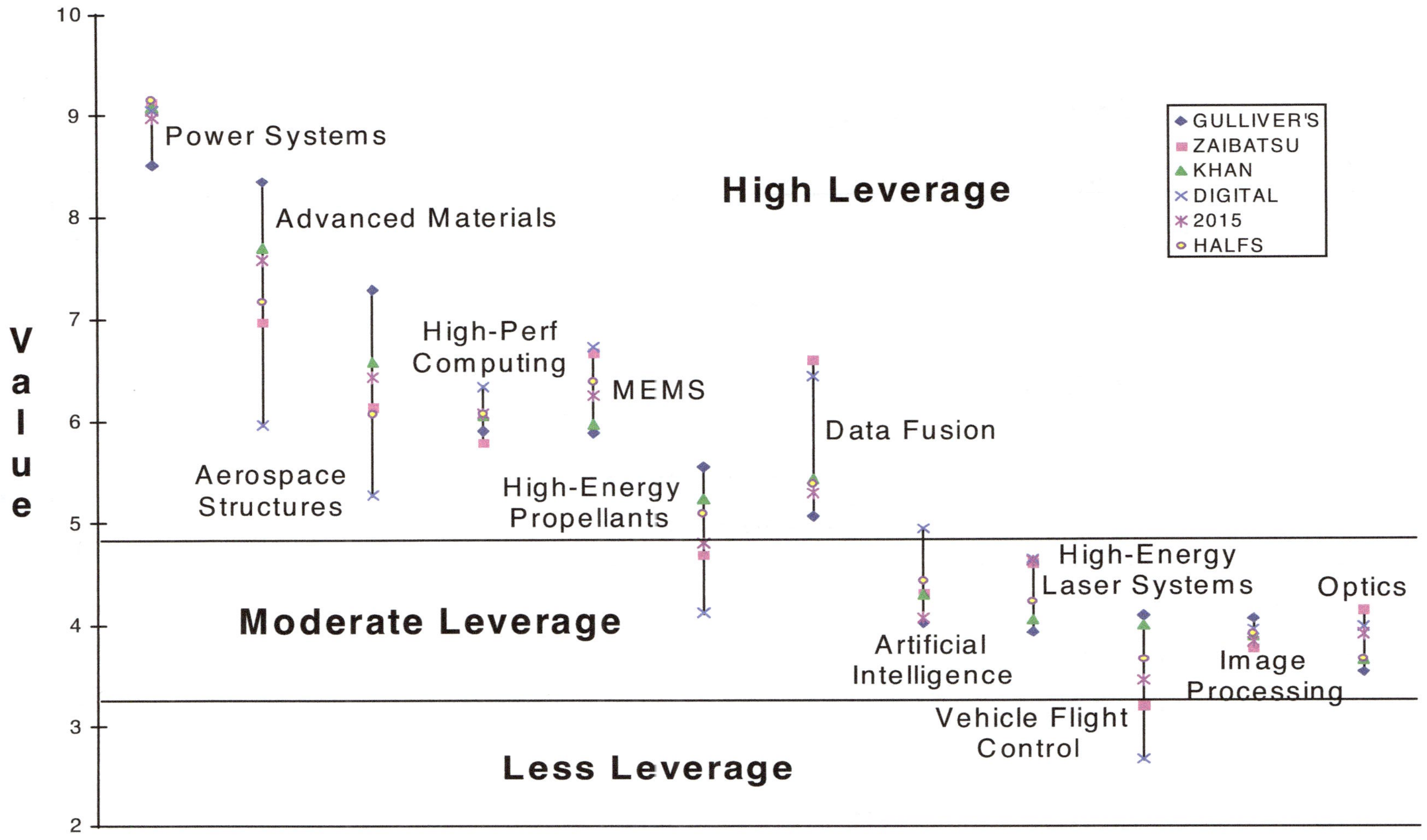

Figure 3-15. Top 12 Technology Rankings (All 43 Sys, Alt Fut Team Wts)

Body-77

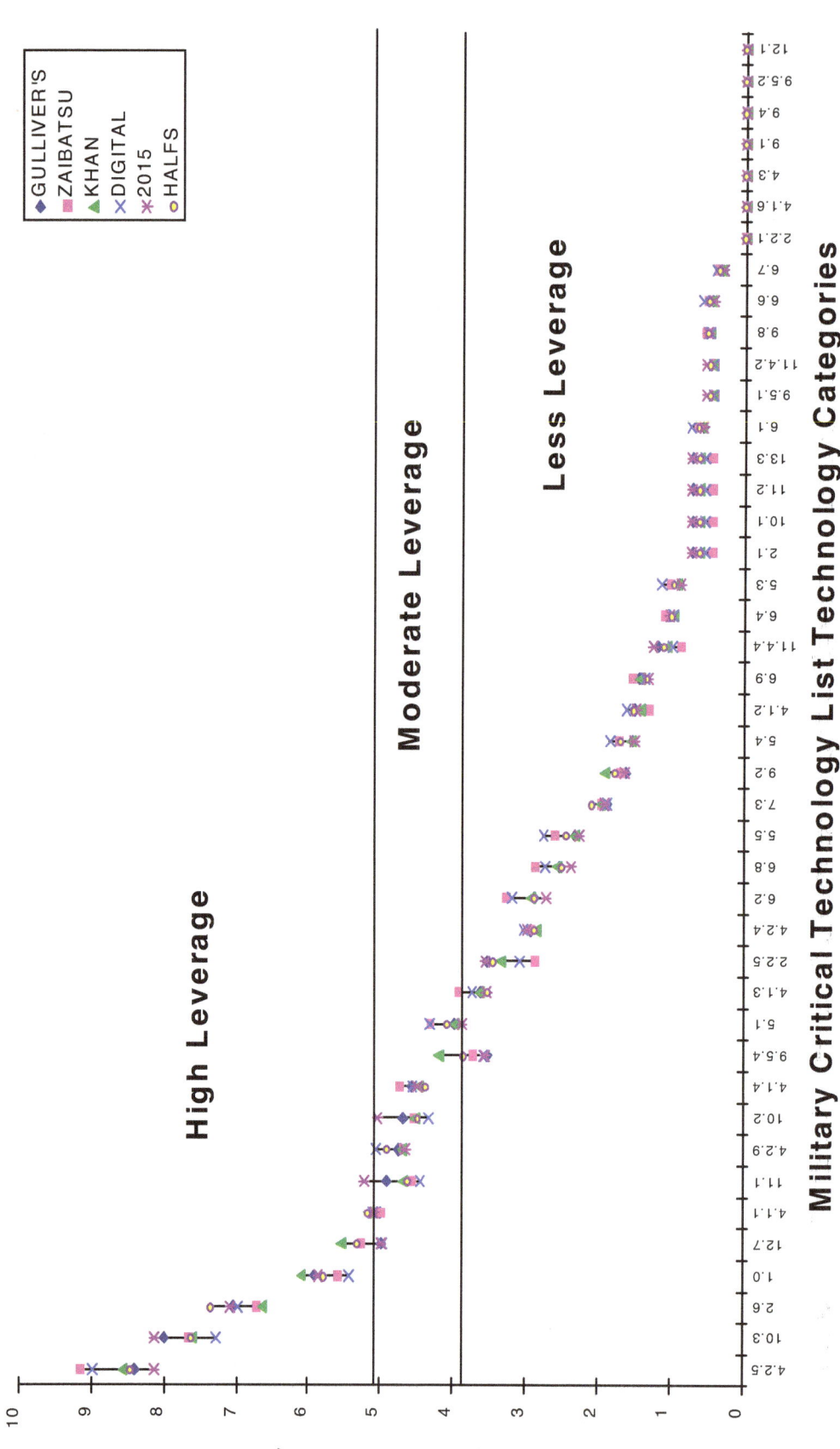

Figure 3-16. Technology Rankings (Top 11 Sys, Alt Fut Team Wts)

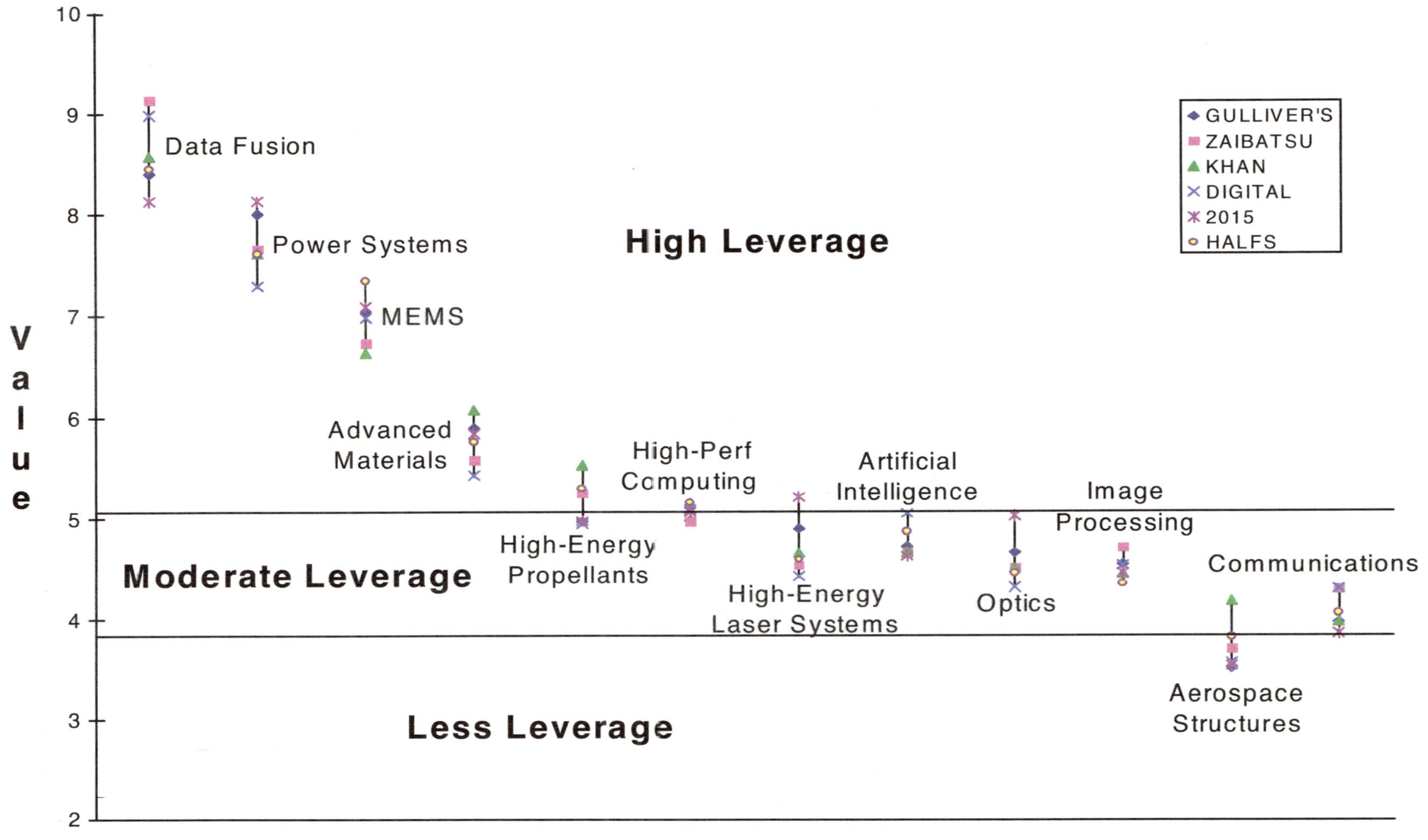

Figure 3-17. Top 12 Technology Rankings (Top 11 Sys, Alt Fut Team Wts)

Body-79

Chapter 4

Conclusions

The *Air Force 2025* Operational Analysis was a milestone in the *Air Force 2025* process and provided a number of unique contributions. Most importantly, it met its fundamental purpose—the OA identified future air and space systems required to dominate combat and the key technologies that will make those systems possible. Further contributions will be covered in the following order:

- the major implications of the study results,
- the lessons learned during the *Air Force 2025* OA process,
- the limitations of the study,
- and the major implications of the *Air Force 2025* OA for the future.

Major Implications of the *Air Force 2025* OA

This analysis contends that the high ground of improved *awareness* offers significant potential for achieving future air and space dominance. Typically, top-scoring systems possessed higher degrees of *awareness* and/or were predominantly *space* systems:

- *Global Information Management System (GIMS)*
- Sanctuary Base (SB)
- Global Surveillance, Reconnaissance, and Targeting System (GSRT)
- Global Area Strike System (GLASS)
- Uninhabited Combat Air Vehicle (UCAV)
- Space-Based Hig-Energy Laser (Space HEL)

- Solar High-Energy Laser (Solar HEL)
- Reconnaissance Unmanned Air Vehicle (Recon UAV)
- Attack Microbots
- Piloted Single Stage Transatmospheric Vehicle (TAV)
- Uninhabited Air-Launched TAV

Seven of the top eight systems emphasized the *awareness* function. *GSRT* can be thought of as a first generation *GIMS*; it obtains most of the value of *GIMS* with much less technological challenge. Both systems scored high because the management of information tasks were assigned high weights by the ***Air Force 2025*** white paper writing teams. Such systems go beyond data fusion to knowledge fusion; they provide a global view that could revolutionize military operations. Improved awareness is critically important because it enables virtually all other air and space force capabilities.

This analysis also suggests that control of the high ground of space will be important. Of the top 11 systems, only three do not operate in space or use major space-based components. Space-based weapons are significant contributors to the operational effectiveness of future air and space operations. They provide key capabilities in space defense, ballistic missile defense, defense of terrestrial forces, and terrestrial power projection. Of the weapon systems evaluated, the Space-Based High-Energy Laser seems to hold the most promise, largely because its optical system could also be used for surveillance and imaging missions (an *awareness* function). Other systems that scored well were the Solar High-Energy Laser, the Space-Based Kinetic Energy Weapon, and the Space-Based High-Powered Microwave. Spacelift is another essential contributor to future space operations (i.e., reusable transatmospheric vehicles provide critical lift capability to improve virtually all space force capabilities).

This analysis also suggests that *improved power* will be best accomplished through improved speed, precision, and on-station time. The ***Air Force 2025*** white paper writing teams viewed the reduction of the observe, orient, decide, act (OODA) loop to a OODA "point" as critical to future operations. All of the "shooter" systems that emphasized *awareness* scored high by reducing the time to identify, target, and kill threats. Among these systems are the GLASS, the Space-Based HEL, and the Solar HEL. The envisioned systems emphasized the increased need for precision over mass, especially with respect to avoiding excess collateral damage. For example, the only system with a weapon of mass destruction was the Asteroid Mitigation System, which incorporated a nuclear device.

The constant quick response requirement of future combat meant many of the systems either were *global* or used uninhabited air vehicles (UAV). It is important to note that while the UAV are uninhabited, none are envisioned as operating autonomously without a human in the loop. Such an improved on-station *power* capability is important because it provides a constant deterrent to enemy forces.

Key to this analysis was the use of several possible alternate futures as the basis for the sensitivity analysis. Because the analysis was conducted across a number of alternate futures and the resulting conclusions remain basically the same across those futures for any reasonable set of weights a future decision maker might apply, these systems are an excellent initial set of systems to consider for future employment of air and space power.

The technology assessment portion of the study identified six high leverage technologies that were are important to a large number of high-scoring systems:

- Data Fusion
- Power Systems
- Micromechanical Devices
- Advanced Materials
- High-Energy Propellants
- High-Performance Computing

Advances in these areas show promise to improve substantially a wide range of air and space operations. Among the technologies, the only surprise to the Analysis team was that *Power Systems* scored so highly. Other technologies were important also, but contributed to only three or four of the high-value systems. Among the top-scoring medium leverage technologies were:

- High-Energy Laser Systems
- Artificial Intelligence
- Optics
- Aerospace Structures
- Image Processing
- Communications

Some of the high leverage technologies enabling *Air Force 2025* systems, including high-performance computing, are being pursued aggressively in the commercial sector. Others, such as data fusion and power systems, have lower commercial value. An expanded analysis of the *Air Force 2025* systems and their embedded technologies can help develop the most effective DOD investment strategy.

OA Process Lessons Learned

Foremost among the *Air Force 2025* OA lessons learned was that the VFT approach worked well. The *Foundations 2025* value model has been used to evaluate systems that span the full range of future air and space combat operations. These systems are conceptual system ideas that will require significant R&D to design and evaluate. The OA provided a structure to incorporate the subjective judgments of operational and technical experts to produce objective, traceable, and robust results.

The focus of the Value Model, *Foundations 2025*, was on the employment of air and space forces. This model does not consider the USAF functional areas required to organize, train, and equip. As it became apparent that none of the current doctrinal frameworks were free of these functional views, the Value Model was developed from the bottom-up. In taking this approach, the Analysis team reduced the institutional biases associated with the numerous stovepipes in the current USAF organizational structure.

Study Limitations

Remember that the analysis did not take into account the cost or risk of developing any of the system concepts. It looked only briefly at the technological challenge of each system concept. While this study indicates some systems and technologies that show promise for dramatically improving the effectiveness of air and space operations, other important factors need to be considered before making an investment decision.

A consequence of most value models is that a complex system (or system of systems) that performs many tasks generally outscores a similar system that performs only a few of the tasks. Also, for *Foundations 2025*, a system's sphere of influence is primarily measured by its range which is only one force quality. For example, the Sanctuary Base scores high because it has awareness, reach, and power capabilities. Yet, it has a 500-mile range limitation on most of those capabilities. *Foundations 2025* would only show a small difference between the Sanctuary Base and a similar system with global range.

Major Implications for the Future

A number of senior decision makers have viewed the model and commented that the best use of **Foundations 2025** may be an analysis of systems within the distinct spheres of *awareness, reach, and power*. They envision separating and developing each function of the model further (refining the tasks, subtasks, force qualities, measures of merit, and scoring functions) and studying which *awareness* (or *reach* or *power*) systems are most promising. These three separate models could be effective mission area analysis tools for the major commands.

The completed **Foundations 2025** value model is the starting point for real value-Focused Thinking. For any function, task, or subtask, the model can be used to evaluate current and projected systems. Next, the acquisition community can *focus* on how new concepts can be developed to significantly increase *value*. Many individual and various creativity techniques can be used to develop these new concepts.

Another opportunity to capitalize on the **Foundations 2025** model is to use it as a framework for future air and space doctrine. Because it identifies fundamental functions, tasks, and subtasks, it could be the foundation for joint doctrine for future air and space warriors. The **Air Force 2025** analysis techniques could be used to develop an entirely new joint military doctrine free from current institutional bias.

Summary

The **Air Force 2025** operational analysis is an important first step. It is offered as a starting point for further discussion and analysis. The **Air Force 2025** process began with creative thinking. The OA completed the process by identifying the most promising systems and the enabling technologies required to provide dominant air and space power for the United States of America as we enter the twenty-first Century.

Appendix A

Foundations 2025 Value Model

This appendix shows the *Foundations 2025* Value Model (fig. A-1 through fig. A-4) and the measures of merit and scoring functions for each of the 134 force qualities (fig. A-5 through fig. A-32), arranged by subtask. It concludes with a glossary of the terms used in the measures of merit and scoring functions.

Value Model

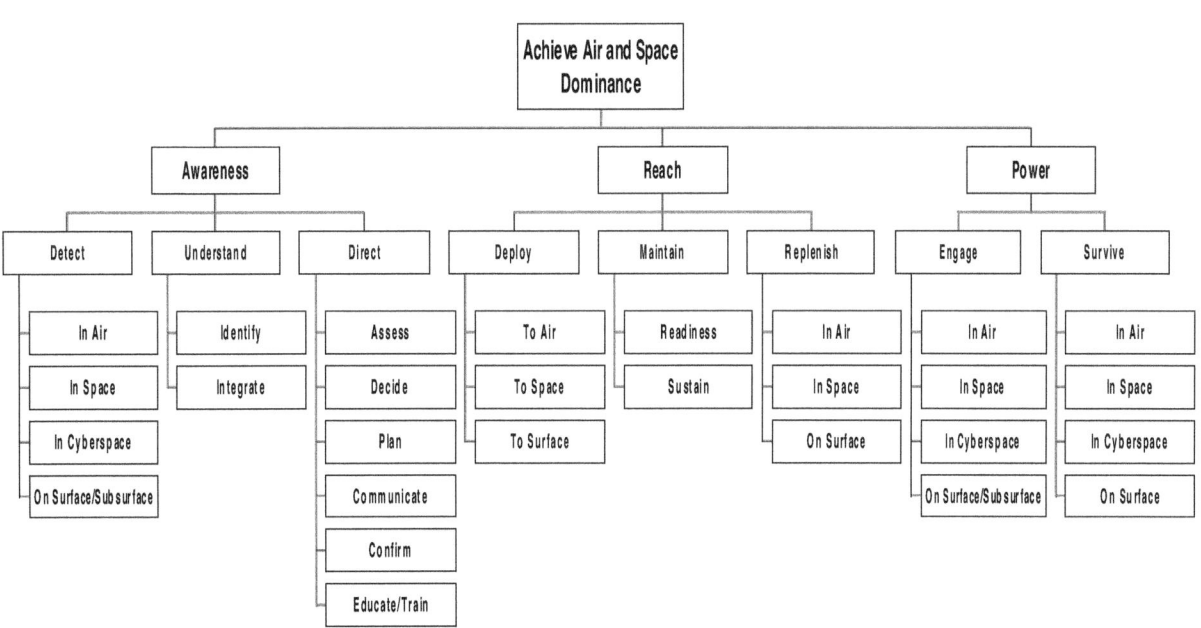

Figure A-1. Value Model - Top Level

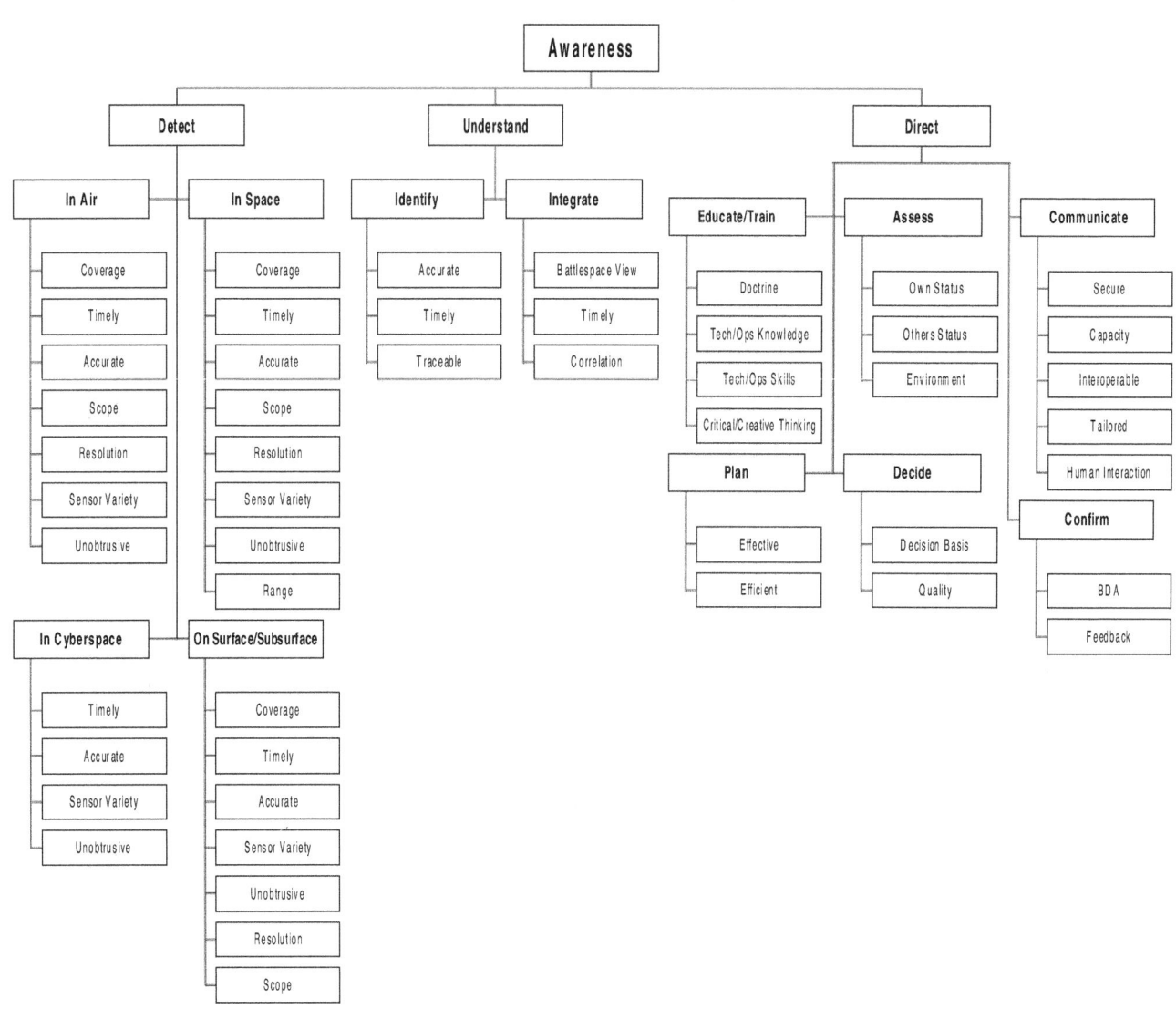

Figure A-2. Value Model - Awareness

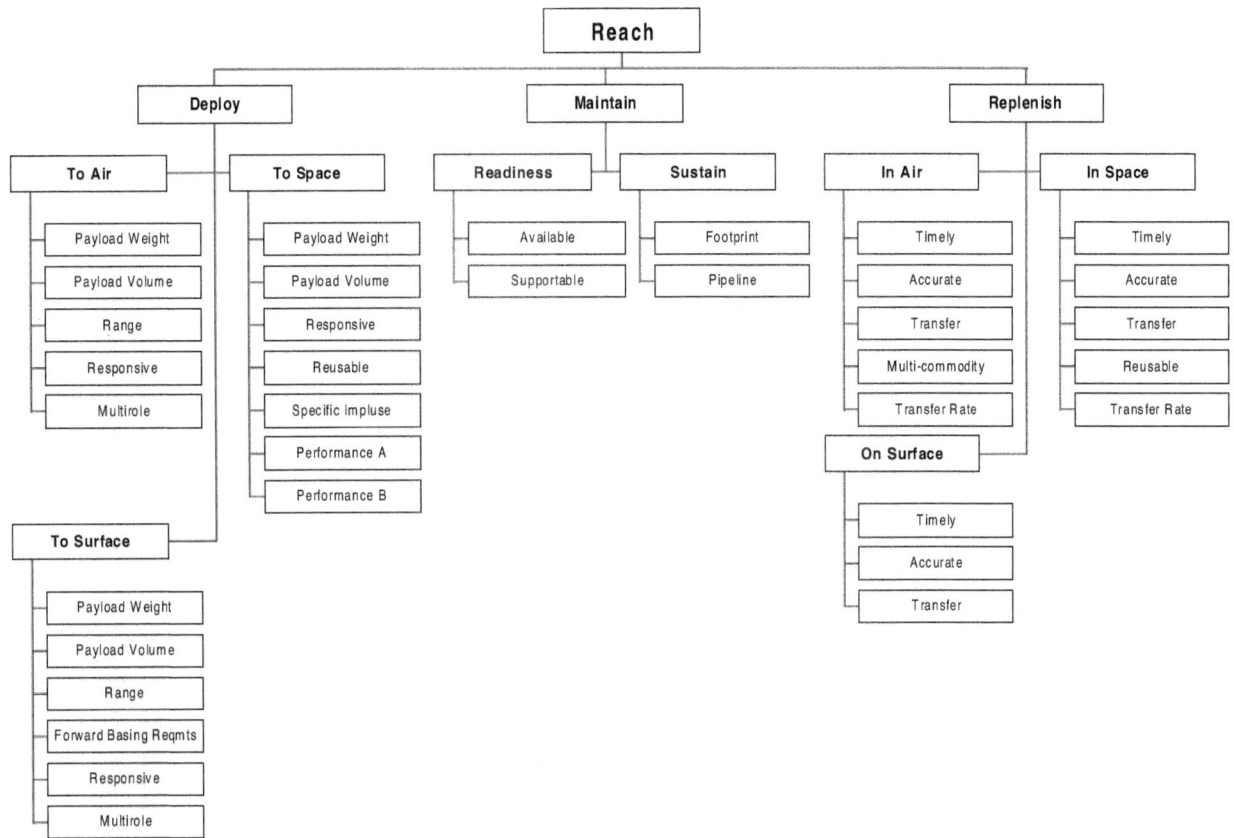

Figure A-3. Value Model - Reach

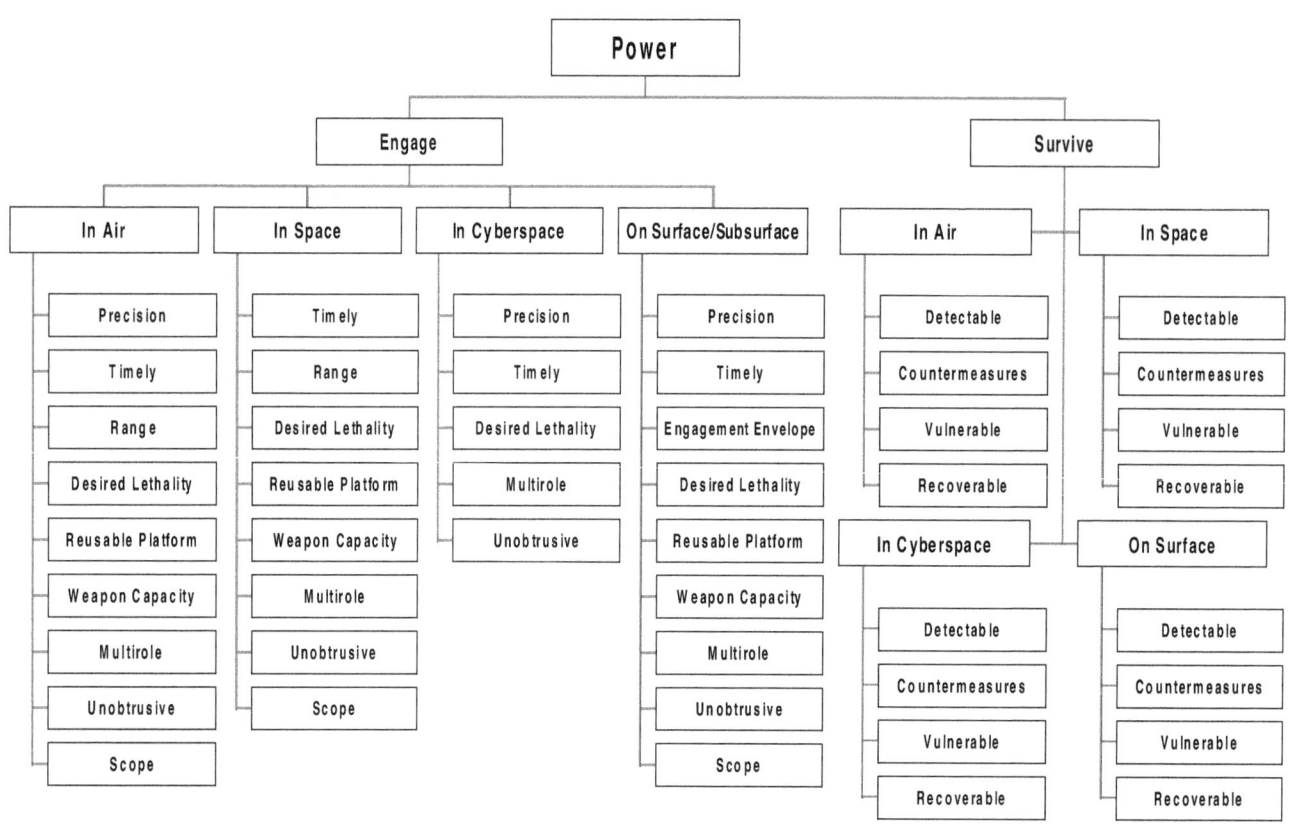

Figure A-4. Value Model - Power

Measures of Merit and Scoring Functions

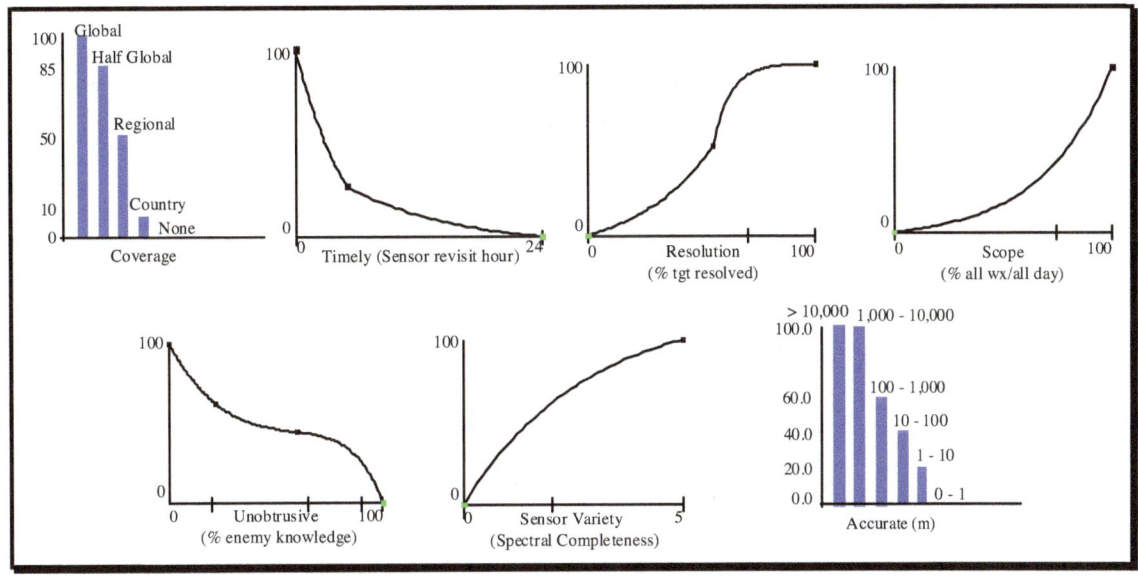

Figure A-5. *Detect in Air* **Scoring Functions**

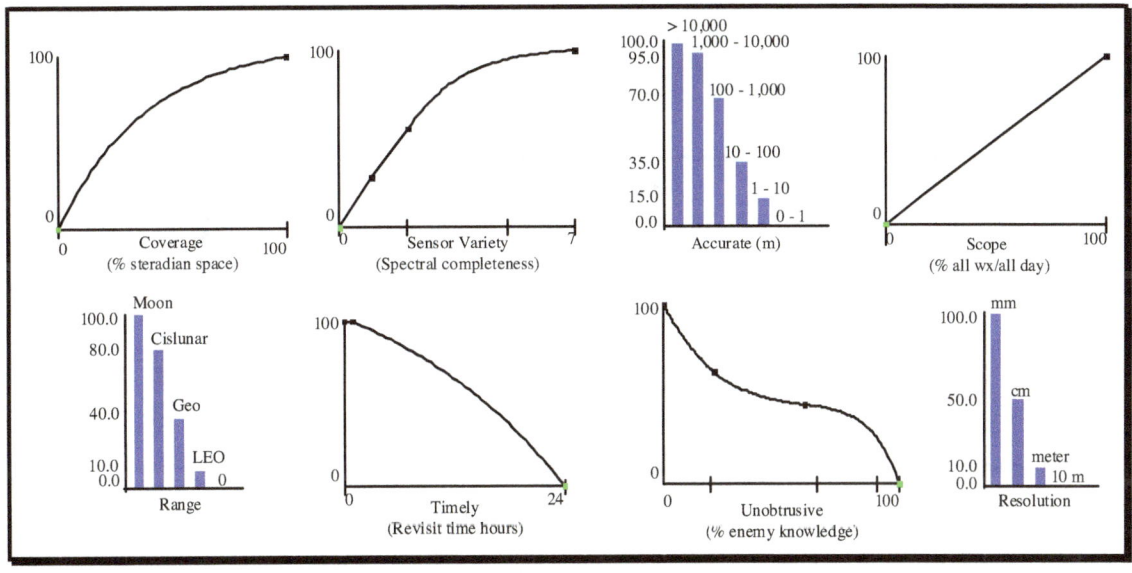

Figure A-6. *Detect in Space* **Scoring Functions**

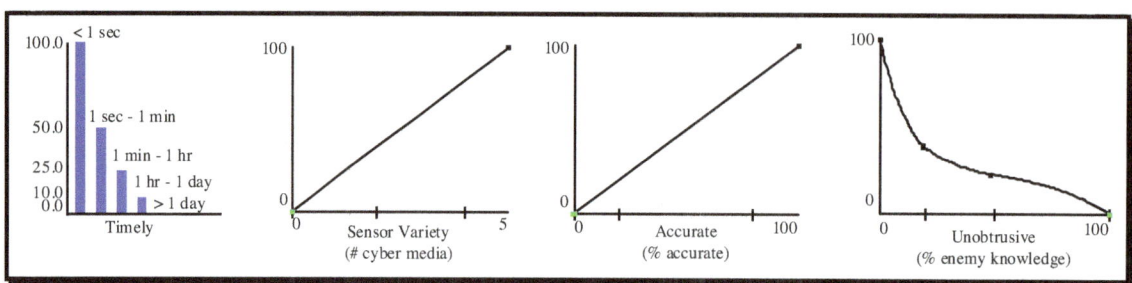

Figure A-7. *Detect in Cyberspace* **Scoring Functions**

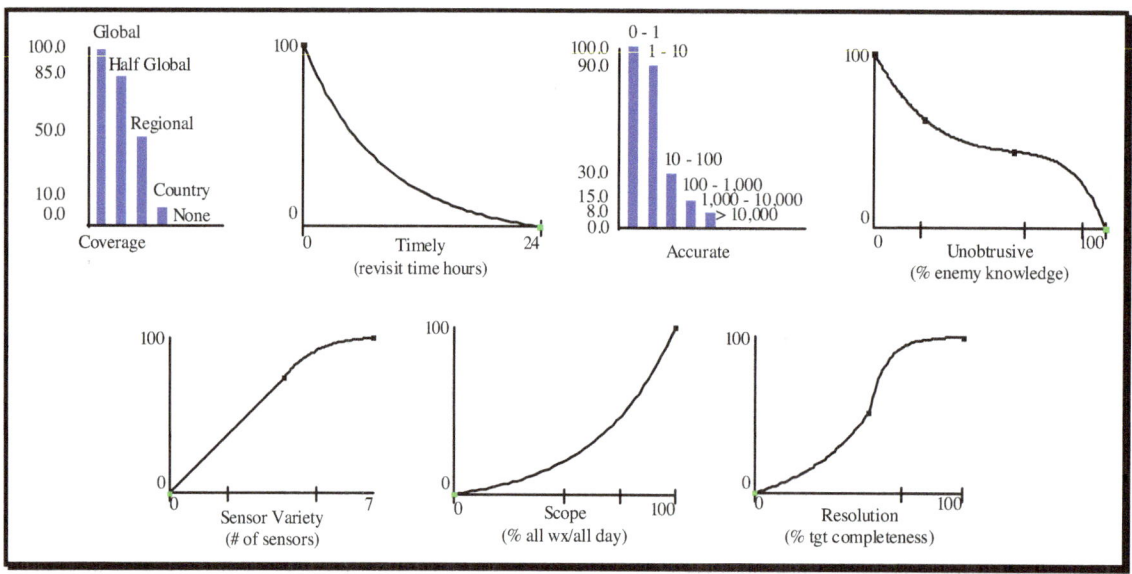

Figure A-8. *Detect on Surface/Subsurface* **Scoring Functions**

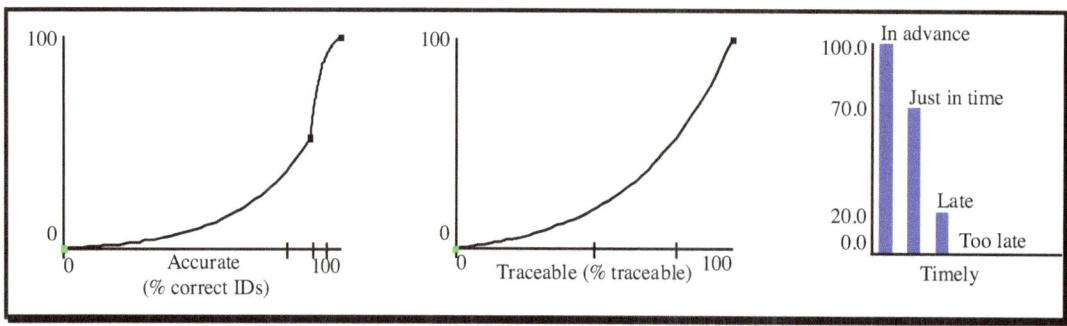

Figure A-9. *Identify* **Scoring Functions**

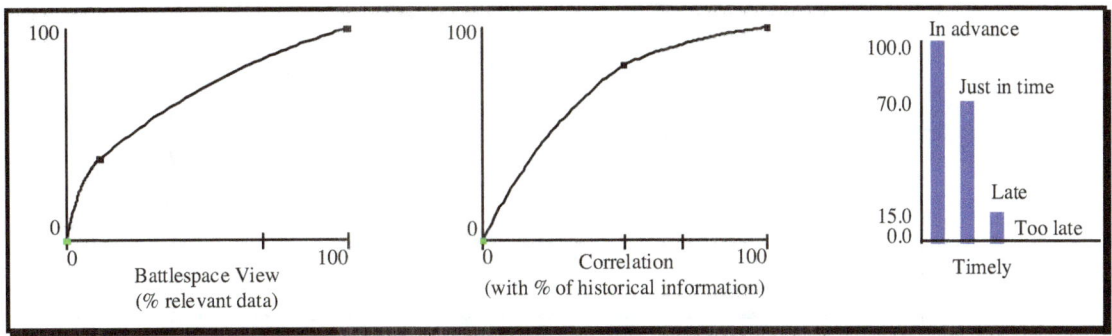

Figure A-10. *Integrate* **Scoring Functions**

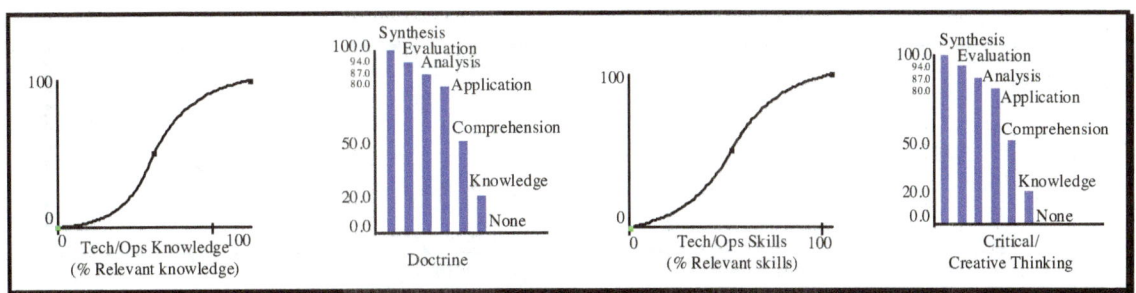

Figure A-11. *Educate/Train* **Scoring Functions**

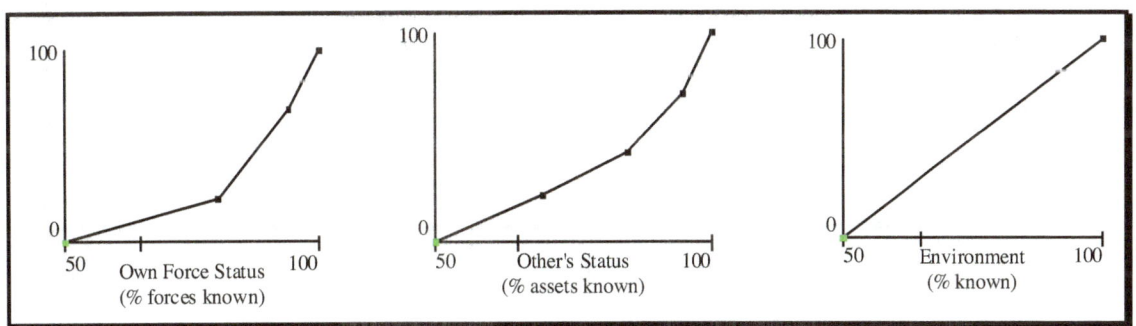

Figure A-12. *Assess* **Scoring Functions**

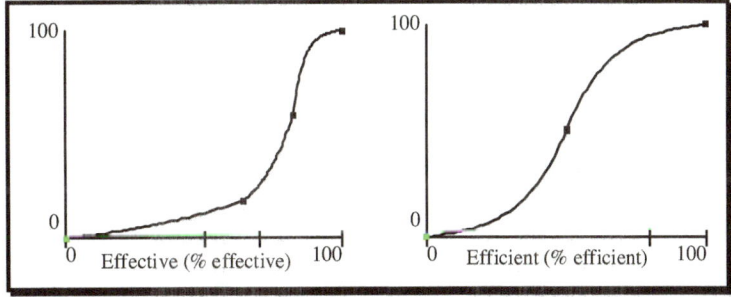

Figure A-13. *Plan* **Scoring Functions**

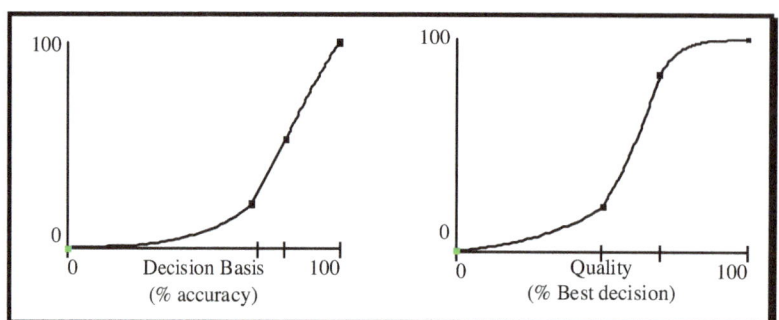

Figure A-14. *Decide* **Scoring Functions**

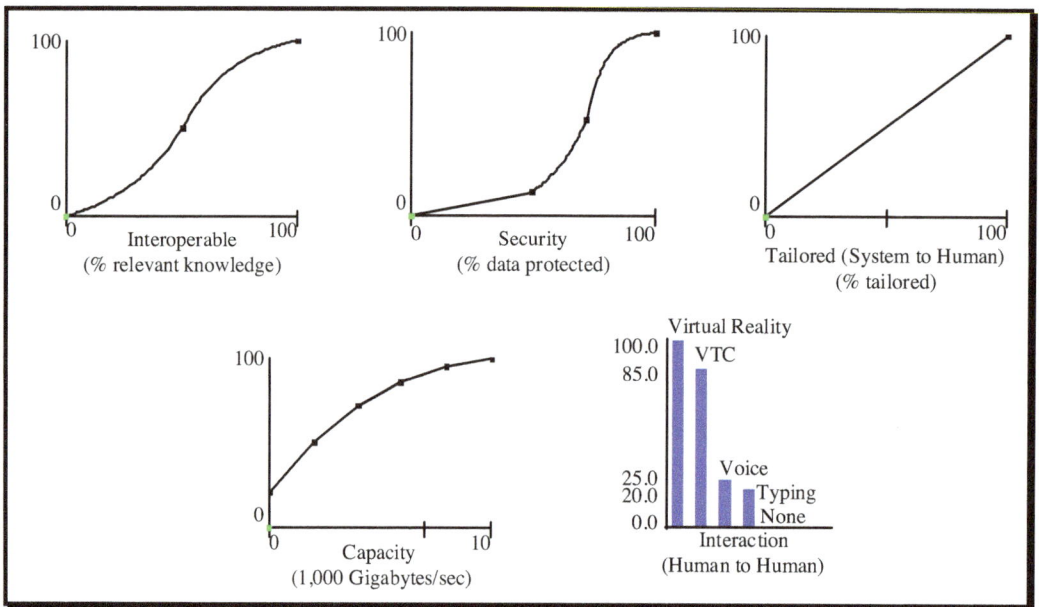

Figure A-15. *Communicate* **Scoring Functions**

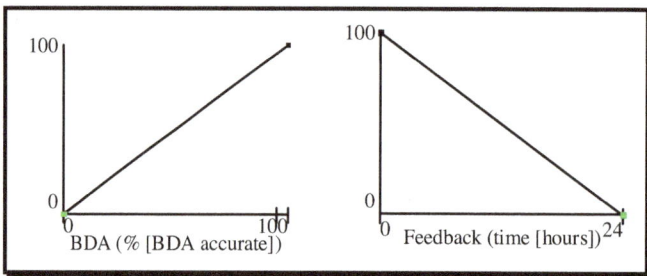

Figure A-16. *Confirm* **Scoring Functions**

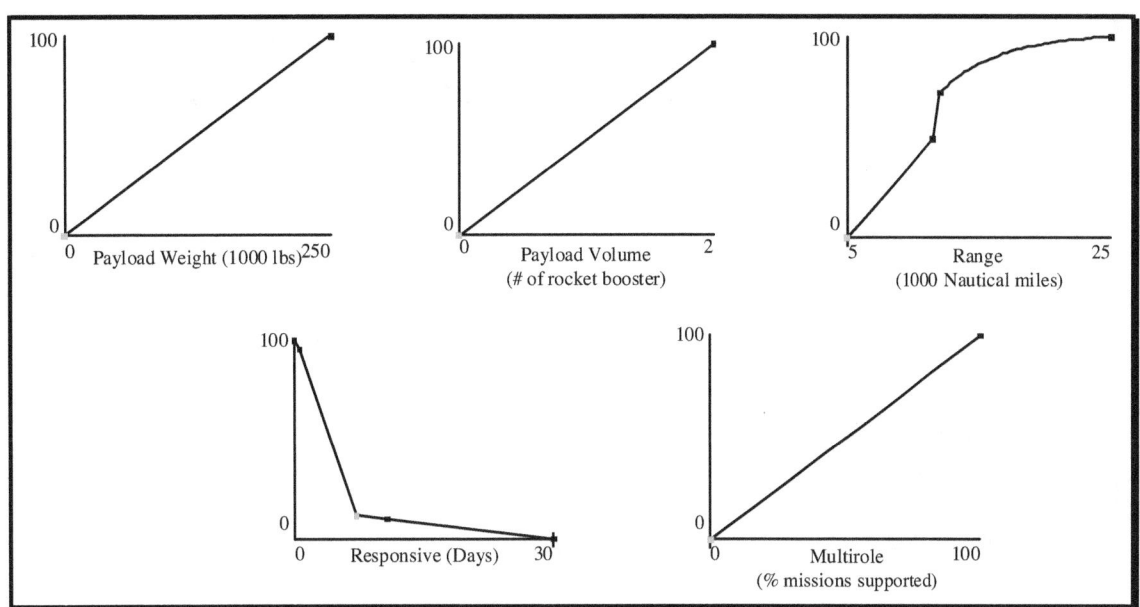

Figure A-17. *Deploy to Air* Scoring Functions

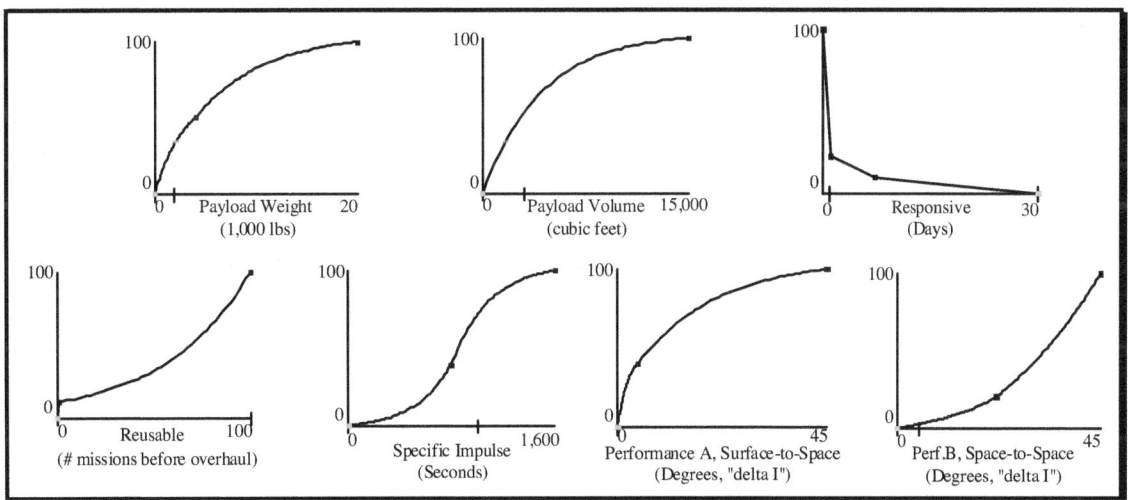

Figure A-18. *Deploy to Space* Scoring Functions

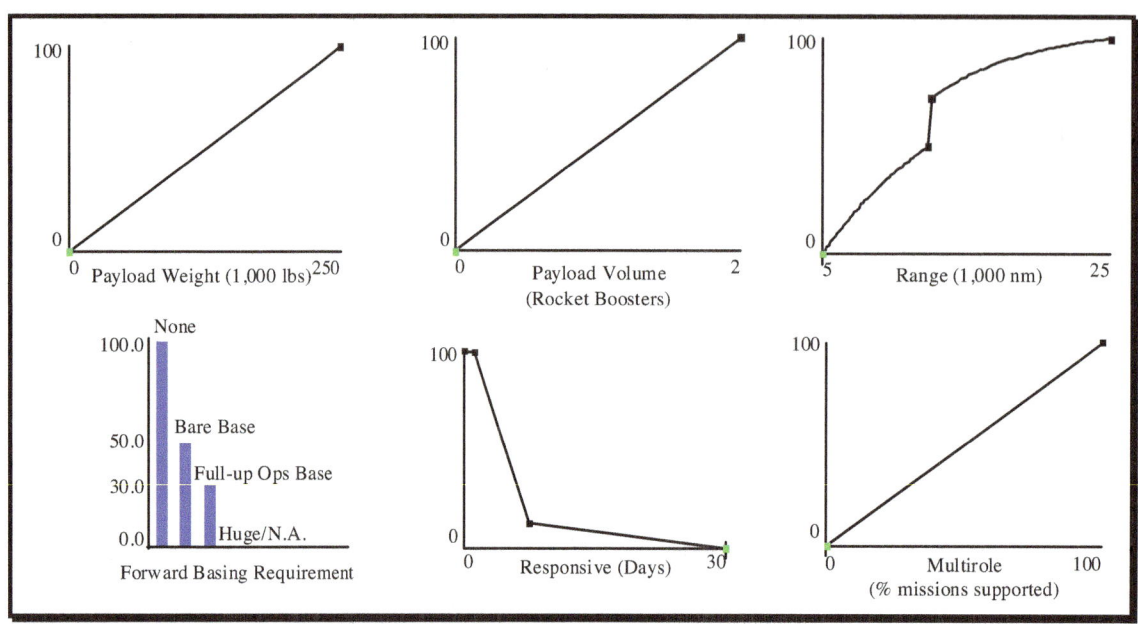

Figure A-19. *Deploy to Surface* Scoring Functions

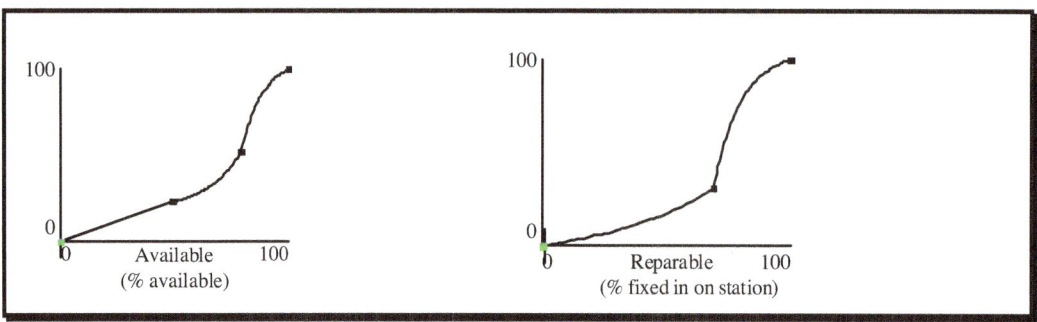

Figure A-20. *Readiness* Scoring Functions

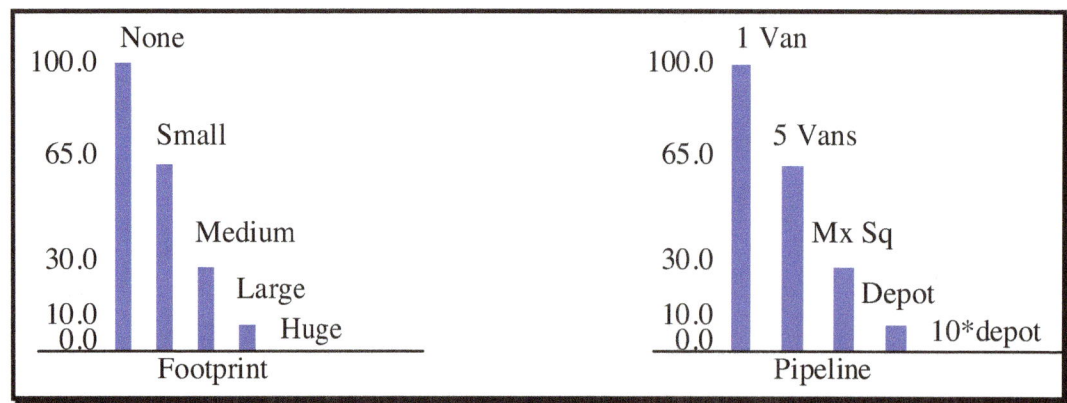

Figure A-21. *Sustain* Scoring Functions

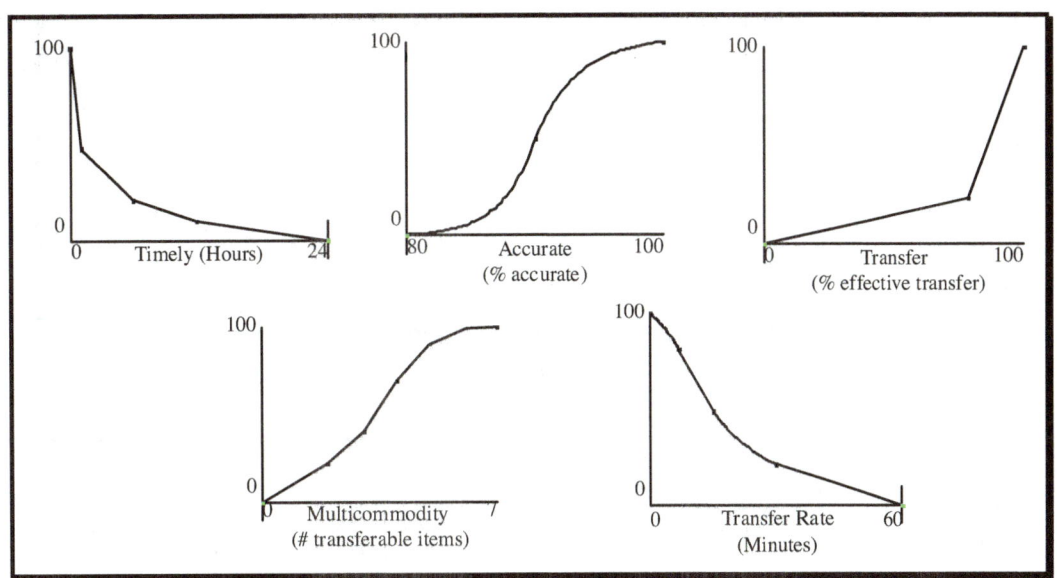

Figure A-22. *Replenish in Air* Scoring Functions

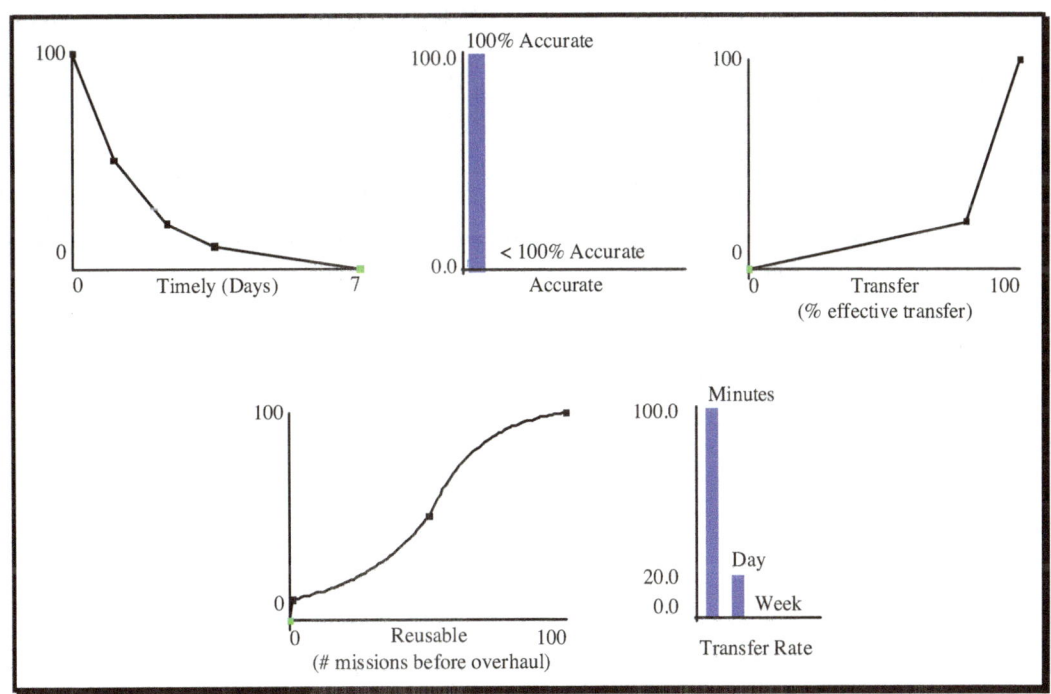

Figure A-23. *Replenish in Space* Scoring Functions

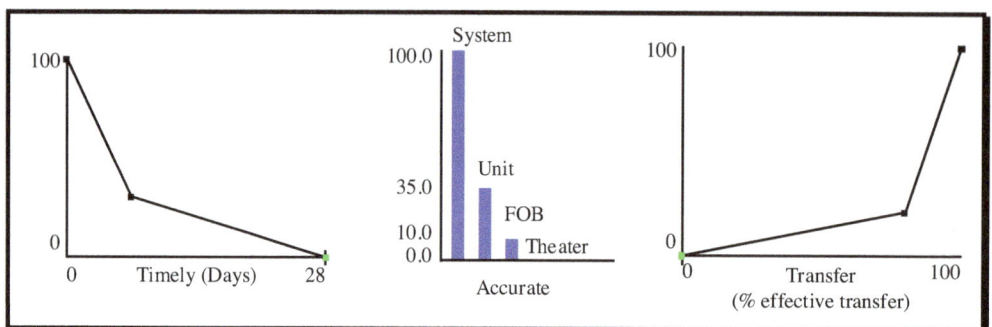

Figure A-24. *Replenish on Surface* **Scoring Functions**

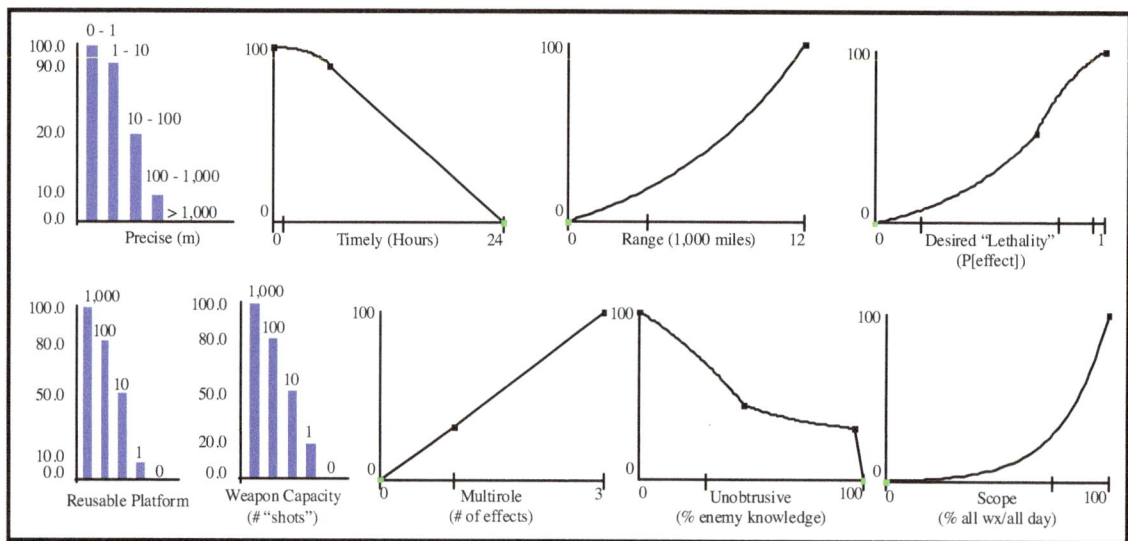

Figure A-25. *Engage in Air* **Scoring Functions**

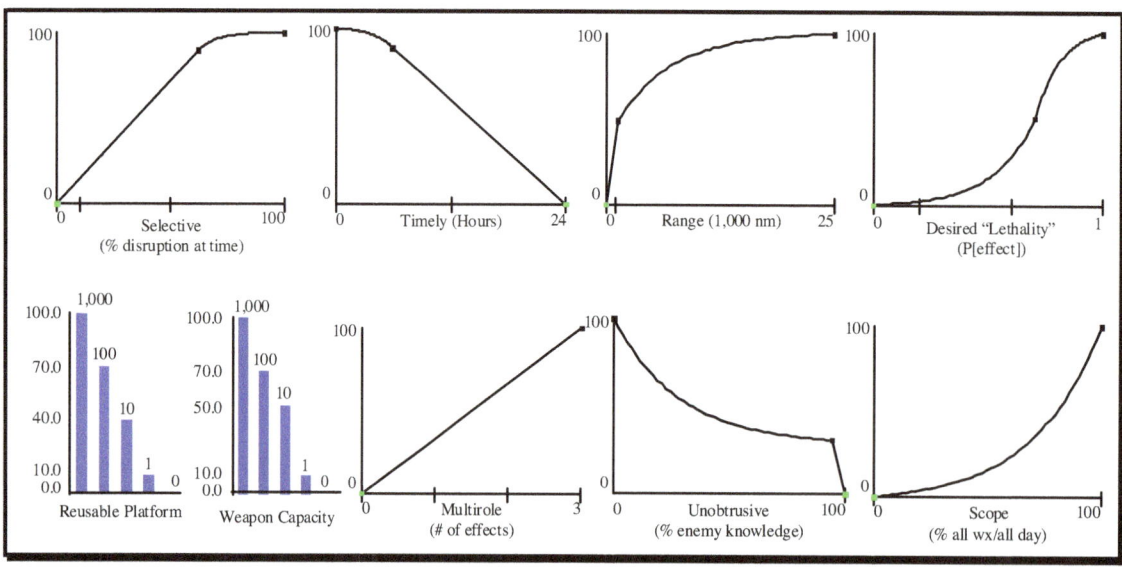

Figure A-26. *Engage in Space* **Scoring Functions**

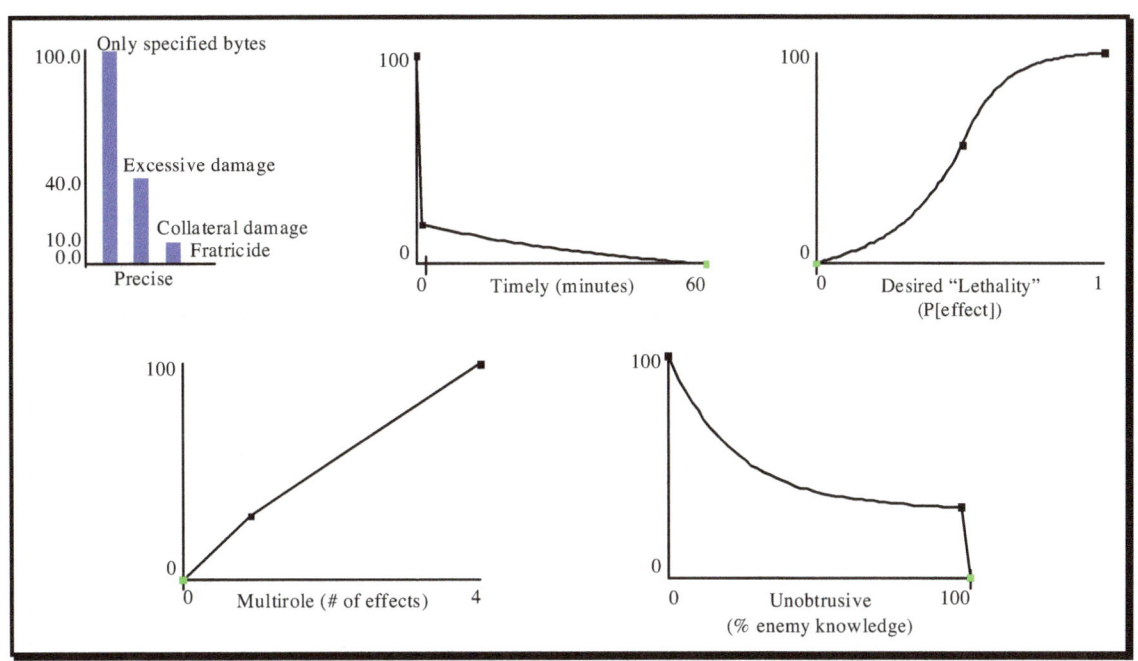

Figure A-27. *Engage in Cyberspace* Scoring Functions

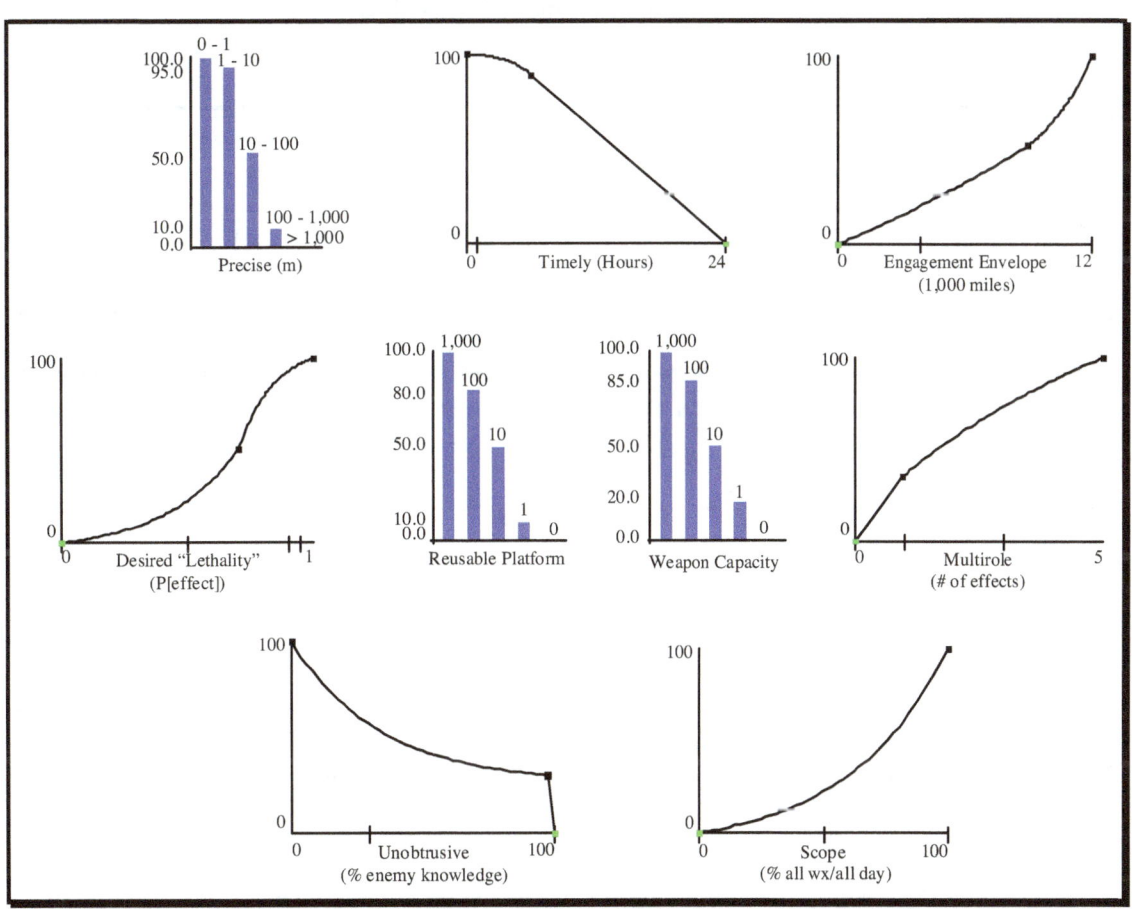

Figure A-28. *Engage on Surface/Subsurface* Scoring Functions

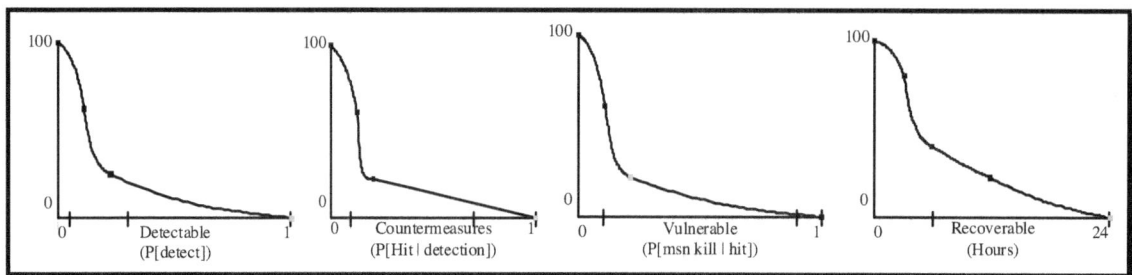

Figure A-29. *Survive in Air* **Scoring Functions**

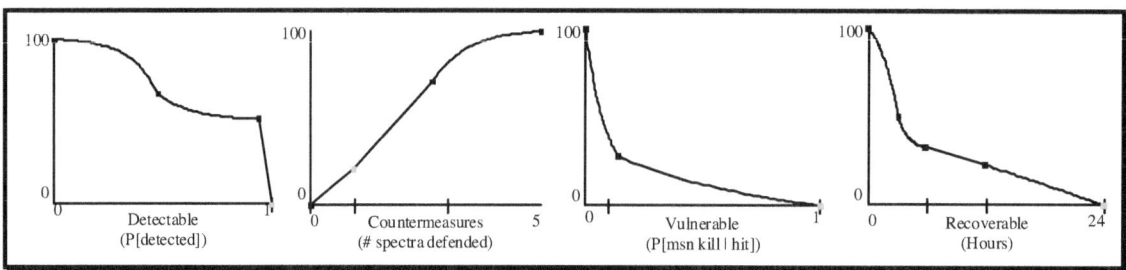

Figure A-30. *Survive in Space* **Scoring Functions**

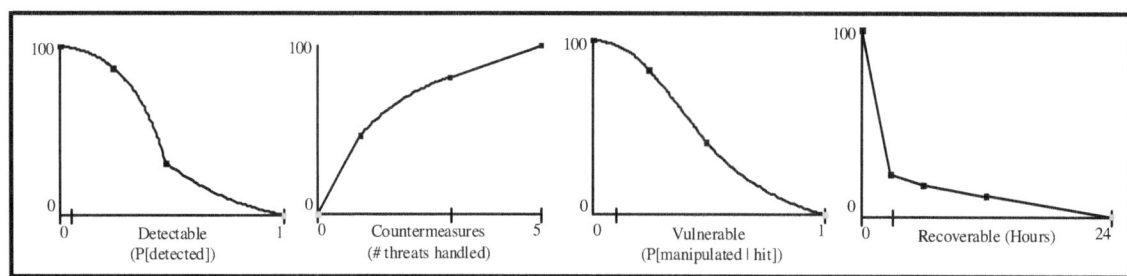

Figure A-31. *Survive in Cyberspace* **Scoring Functions**

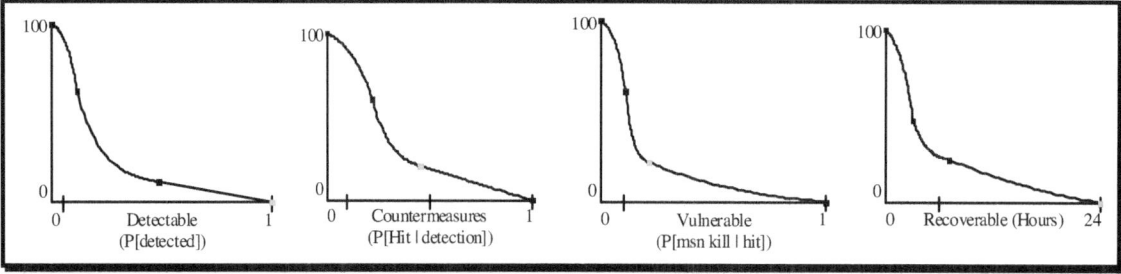

Figure A-32. *Survive on Surface* **Scoring Functions**

Operational Analysis Glossary

Air - 1. b. This mixture with varying amounts of moisture, low-altitude pollutants, and particulate matter, enveloping the earth: atmosphere.[1]

[Awareness] Aware - 1. Having knowledge or cognizance. 2. Vigilant: watchful.[2]

Appreciate - 1. To recognize the quality, significance, or magnitude of.[3]

Assess - 1. To estimate the value of 4. To appraise or evaluate.[4]

Command - 1. To give orders to: direct. 2. To have authoritative control over: rule.[5]

Command - 2. An order given by a commander; that is, the will of the commander expressed for the purpose of bringing about a particular action.[6]

Communicate - 1. To have an interchange, as of ideas or information. 2. To express oneself effectively.[7]

Communications - A method or means of conveying information of any kind from one person or place to another.[8]

Control - 1. To exercise authority or influence over: direct.[9]

Control - 1. Authority which may be less than full command exercised by a commander over part of the activities of subordinate or other organizations.[10]

Cyberspace - The virtual space of computer memory and networks, telecommunications and digital media.[11]

Decide - 1. To pronounce a judgment. 2. To make up one's mind.[12]

Decision - In an estimate of the situation, a clear and concise statement of the line of action intended to be followed by the commander as the one most favorable to the successful accomplishment of his mission.[13]

Deploy - 1. To station (persons or forces) systematically over an area.[14]

Deployment - 3. In a strategic sense, the relocation of forces to desired areas of operation.[15]

Detect - 1. To discover or discern the existence, presence, or fact of.[16]

Detection - The discovery by any means of the presence of a person, object, or phenomenon of potential military significance.[17]

Educate - 1. a. To provide with training or knowledge, esp. via formal schooling: teach. b. To provide with training for a specific purpose, as a vocation. 2. To provide with information: inform. 3. To stimulate or develop the mental or moral growth of.[18]

Employ - 1. To put to service or use. 2. To apply or devote (e.g., time) to an activity. 3. a. To put to work.[19]

Engage - 1. To obtain or contract for the services of: employ. 2. To contract for the use of: reserve. 3. To obtain and hold the attention of: engross. 4. To require the use of: occupy. 6. To enter or bring into conflict with. To interlock or cause to interlock: mesh. 8. To win: attract. 9. To involve: entangle.[20]

Engage - In air intercept, a code meaning, "Attack designated contact."[21]

Global - 2. Of, pertaining to, or involving the entire earth: worldwide. 3. Total: comprehensive.[22]

Global Awareness - The ability to use affordable means to derive appropriate information about one or more places of interest after a delay which is short enough to satisfy operational needs.[23]

Identify - 1. b. To find out the origin, nature, or definitive elements of.[24]

Identification - 1. The process of determining the friendly or hostile character of an unknown detected contact.[25]

Imbue - 2. To inspire, permeate, or pervade.[26]

Integrate - 1. To make into a whole by bringing all parts together: unify. 2. To join with something else: unity.[27]

Maintain - 2. To preserve or keep in a given existing condition, as of efficiency or good repair. 3. a. To provide for. b. To keep in existence: sustain.[28]

Maintenance - 1. All action taken to retain materiel in or to restore it to a specified condition. It includes: inspection, testing, servicing, classification as to serviceability, repair, rebuilding, and reclamation. 2. All supply and repair action taken to keep a force in condition to carry out its mission. 3. The routine recurring work required to keep a facility (plant, building, structure, ground facility, utility system, or other real property) in such condition that it may be continuously utilized, at its original or designed capacity and efficiency, for its intended purpose.[29]

Plan - 1. To formulate a scheme or program for the accomplishment or attainment of.[30]

Power - 1. The ability or capacity to act or perform effectively. 3. Strength or force exerted or capable of being exerted: might. 4. The ability or official capacity to exercise control over others. 5. A person, group, or nation having great influence or control over others. 6. The might of a nation, political organization, or similar group.[31]

Prepare - To make ready in advance for a particular purpose, event, or occasion.[32]

Reach - 1. The act or power of stretching or thrusting out. 2. The distance or extent something can reach. 3. b. The range or scope of influence or effect.[33]

Replenish - To fill or make complete again: add a new supply to.[34]

Space - 2. The expanse in which the solar system, stars, and galaxies exist: universe.[35]

Subsurface - Below, under, beneath the surface.[36]

Surface - 2. b. The two-dimensional locus of points located in three-dimensional space whose height z above each point (x,y) of a region of a coordinate plane is specified by a function $f(x,y)$ of two arguments.[37]

Survive - 1. To remain alive or in existence. 2. To persist through.[38]

Train - 1. To coach in or accustom to a mode of behavior or performance. 2. To make proficient with special instruction and practice.[39]

Understand - 1. To perceive and comprehend the nature and significance of.[40]

Notes

[1] *Webster's II New Riverside University Dictionary.*
[2] Ibid.
[3] Ibid.
[4] Ibid.
[5] Ibid.
[6] DOD; IADB; and *JCS Pub 1.*
[7] *Webster's Dictionary.*
[8] DOD; IADB; and *JCS Pub 1.*
[9] *Webster's Dictionary.*
[10] DOD; IADB; and *JCS Pub 1.*
[11] Cotton, Bob, *The Cyberspace Lexicon: An Illustrated Dictionary of Terms From Multimedia to Virtual Reality* (London: Phaedon, 1994).
[12] *Webster's Dictionary.*
[13] DOD; IADB; and *JCS Pub 1.*
[14] *Webster's Dictionary*
[15] NATO; DOD; IADB; and *JCS Pub 1.*
[16] *Webster's Dictionary.*
[17] NATO; and *JCS Pub 1.*
[18] *Webster's Dictionary.*
[19] Ibid.
[20] Ibid.
[21] DOD; and *JCS Pub 1.*
[22] *Webster's Dictionary.*
[23] USAF Scientific Advisory Board, *New World Vistas: Air and Space Power for the 21st Century,* summary volume (Washington, D.C.: USAF Scientific Advisory Board, 15 December 1995).
[24] *Webster's Dictionary.*
[25] DOD; IADB; and *JCS Pub 1.*
[26] *Webster's Dictionary.*
[27] Ibid.
[28] Ibid.
[29] NATO; and *JCS Pub 1.*
[30] *Webster's Dictionary.*
[31] Ibid.
[32] Ibid.
[33] Ibid.
[34] Ibid.

[35] Ibid.
[36] Ibid.
[37] Ibid.
[38] Ibid.
[39] Ibid.
[40] Ibid.

Appendix B

System Descriptions

This appendix provides a description of each of the 43 systems identified in the *AF 2025* study. Figure B-1 shows the system hierarchy, categorized by functional area. Each system description contains a brief narrative, a list of capabilities, enabling technologies, and *AF 2025* white papers relating to the system. Table 4 lists the *AF 2025* student writing teams.

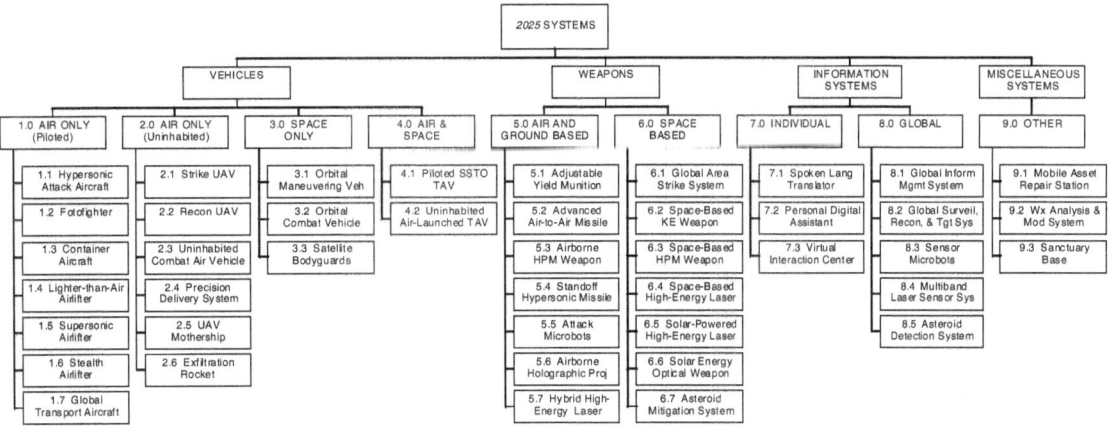

Figure B-1. System Hierarchy

Table 4

Writing Team Letter Designators and Names

Letter	Name
A	Counterair
B	Planetary Defense
C	Counterspace
D	Operability and Defense
E	Counterinformation
F	Airlift
G	Spacelift
H	Aerospace Replenishment
I	Frontier Missions
J	Close Air Support
K	Interdiction
L	Strategic Aerospace Warfare Study
M	Information Operations
N	Strategic Attack
O	C^2 Attack
P	Strategic & C^2 Attack
Q	Special Operations
R	Special Operations
S	RPV & Aerospace
T	Space Surveillance & Reconnaissance Fusion
U	Surveillance & Reconnaissance Information Operations
V	Surveillance & Reconnaissance Real Time Integration
W	On-Orbit Support
X	General Education & Training
Y	Training & Readiness
Z	Information Technology Integration in Education & Training
β	USAFA Information Technology Management
Δ	Space Operations
Φ	Logistics
Γ	Weather Modification
Λ	Alternate Futures
Π	AFIT Operational Analysis
Θ	Combat Support
Σ1	AFIT Space Operations
Σ2	AFIT Space Operations
Ω	AFIT Logistics
Ψ1	USAFA Hypersonics
Ψ2	USAFA Hypersonics
Ψ3	USAFA Hypersonics

1.1 Hypersonic Attack Aircraft

Brief Description

A high-speed strike vehicle capable of projecting lethal force anywhere in the world in less than four hours. Operating at Mach 12 and a cruise altitude of 100,000 feet, this vehicle is a reusable two-stage system comprised of an unmanned boost vehicle and a manned hypersonic strike aircraft. The gas turbine boost vehicle requires a conventional runway and accelerates the strike vehicle to Mach 3.5 and 65,000 ft. The strike vehicle then separates and uses a Ramjet/supersonic combustion Ramjet (Scramjet) engine to reach its cruise condition. The total system range is 10,000 NM; the hypersonic strike vehicle has an unrefueled range of 5,000 NM. It is capable of launching precision guided munitions, including the hypersonic air-to-ground missile described in system 5.4, at a standoff distance of 1,450 NM. Alternatively, the platform may be used to transport an uninhabited UAV described in system 4.2.

Capabilities

- Weapons suite includes precision guided munitions including up to 10 standoff hypersonic missiles
- Designed to transport an uninhabited TAV to 100,000 ft
- Requires turbine-engined supersonic launch platform to achieve operating condition

Enabling Technologies (MCTL)

- 1.0, Materials
- 4.1, Digital Processing
    - 4.1.1, High-Performance Computing
- 9.2, Ramjet, Scramjet, and Combined Cycle Engines
- 9.5.4, Aerospace Structures and Systems
- 12.7, Energetic Materials

White Papers

- Ψ, Hypersonics
-

1.2 Fotofighter

Brief Description

A highly maneuverable, stealthy, inhabited advanced fighter aircraft whose skin is fitted with an array of diode lasers and sensors. Efficient electronic control of the laser arrays allows this fighter to engage multiple targets simultaneously with varying degrees of lethality. At low powers, the arrays can function as transmitters and receivers for low probability of intercept (LPI) communications. Threat detection, target illumination, and tracking are also possible.

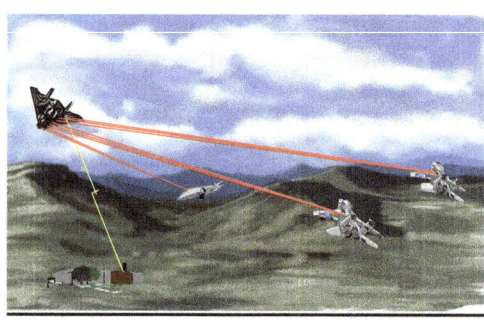

Capabilities

- Automatic threat detection, tracking, and target illumination
- Transmitter and receiver for LPI communications
- Energy replenishment for directed energy weapons (DEW) armament from other air/ground DEW platforms while airborne

Enabling Technologies (MCTL)

- 1.0, Materials
- 2.6, Micromechanical Devices (MEMS)
- 9.5, Aerospace Structures and Systems
- 4.1, Digital Computing
    - 4.1.1, High-Performance Computing
- 4.2.5, Data Fusion
- 4.3, Hybrid Computing
- 6.0, Sensors and Electronic Combat Technology
    - 6.2, Optical Sensors
    - 6.8, Radar Sensors
- 10.1, Lasers
- 11.1, High-Energy Laser (HEL) Systems

White Papers

- A, Counter Air

1.3 Container Aircraft

Brief Description

The container aircraft consists of an airlifter in which standard shipping containers would form integral structures of the fuselage. The aircraft consists of three baseline sections: (1) the cockpit, (2) the wingbox, and (3) the empennage. In its simplest form, the "short" version, the aircraft is capable of flight by joining the cockpit, wingbox, and empennage directly together. With standard shipping containers installed between the cockpit and wingbox, the wingbox and the empennage, the aircraft would be configured to carry cargo ("stretch" version). The first wave of container aircraft to arrive in a theater of operations would be "disassembled." The cockpit would form a command and control facility, the aircraft engines would generate the base power, the wings could provide fuel storage, and the containers themselves (when empty) would provide shelter for troops, supplies, and equipment. This concept will provide a mobile base.

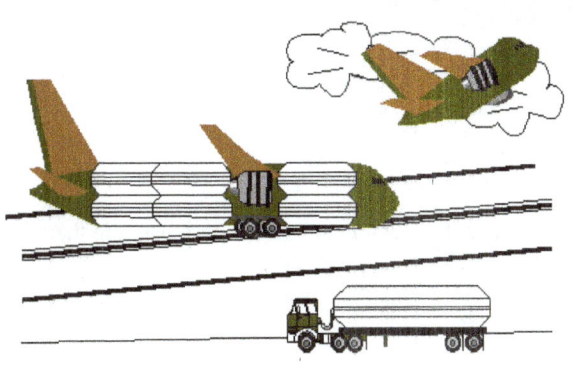

Capabilities

- Modular aircraft using "standard-ized" modular cargo containers as integral fuselage components
- Provides intertheater airlift of troops, supplies, equipment and MEDVAC
- When disassembled at a bare base, provides mobile base facilities
- C^3I center (cockpit), power station (engines), fuel storage (wings), base buildings (empty containers)

Enabling Technologies (MCTL)

- 1.0, Materials
- 7.3, Vehicle and Flight Control
- 9.5, Aerospace Structures and Control
    - 9.5.4, Aircraft High-Performance Structures

White Papers

- Ω, Logistics

1.4 Lighter-than-Air Airlifter

Brief Description

A large capacity, rigid-frame, lighter-than-air vehicle that provides 1-million-pound airlift capability with a unrefueled range of 12,500 NM. This vehicle would also have the ability to deploy and recover powered UAVs while stationary or in-transit. Vehicle is able to house support materiel, personnel and MEDVAC modules depending upon mission requirements.

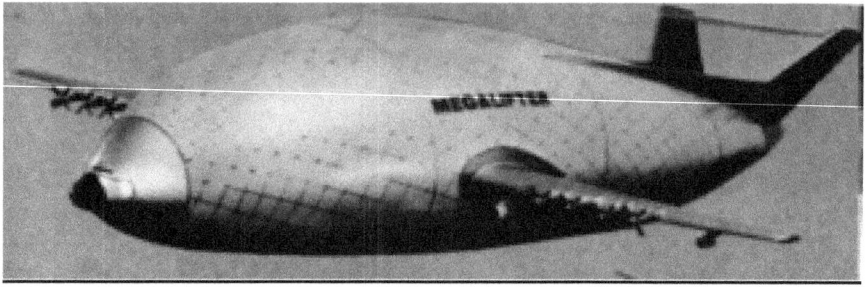

Capabilities

- Global transport of 1 million pounds of materiel, equipment and personnel to/from bare bases in theater
- Deploy and recover UAVs while airborne
- Transport and house MEDEVAC modules

Enabling Technologies (MCTL)

- 1.0, Materials
- 4.1, Digital Processing
    - 4.1.1, High-Performance Computing
- 7.3, Vehicle and Flight Control
- 9.5.4, Aircraft High-Performance Structures

White Papers

- F, Airlift

1.5 Supersonic Airlifter

Brief Description

A Mach 2.4 supersonic airlifter that provides 50,000-pound airlift capability with a unrefueled range of 5,000 NM. This vehicle would provide the capability to deliver military personnel (roughly 150), advanced precision weapons, and appropriate resupply world wide within hours.

Capabilities

- Rapid global transport of personnel and support materiel
- Mach 2.4 cruise speed with a payload of 50,000 pounds; unrefueled range of 5,000 NM

Enabling Technologies (MCTL)

- 1.0, Materials
- 2.6, Micromechanical Devices (MEMS)
- 9.5.4, Aircraft High-Performance Structures
- 9.1, Gas Turbine Propulsion Systems
- 4.1, Digital Processing
    - 4.1.1, High-Performance Computing

White Papers

- F, Airlift

1.6 Stealth Airlifter

Brief Description

The stealth airlifter (SA) is an all-weather, low observable, aircraft capable of low- supersonic cruise and dedicated to special operations forces (SOF). With an unrefueled range up to 4,000 NM, it would be used for insertion and extraction of SOF teams, as well as for extraction of high-value assets (HVA) and weapons of mass destruction. The SA would be connected to a global information management system(e,g., GIMS System 8.1), for all source intelligence, weather, navigation, and communications.

Capabilities

- Capable of operating from in-theater operating bases; all-weather, low observable, supersonic cruise aircraft designed to support SOF missions
- Vertical/short take off and landing (V/STOL) capable
- Can transport 2,500 pounds up to 4,000 NM

Enabling Technologies (MCTL)

- 1.0, Materials
- 9.1, Gas Turbine Propulsion Systems
- 9.5.4, Aircraft High-Performance Structures

White Papers

- R, Special Operations

1.7 Global Transport Aircraft

Brief Description

A global reach transport airplane of less than 1 million pounds take-off gross weight, capable of carrying 150,000-250,000 pounds 12,000 to 10,000 NM, respectively. This vehicle also would have the ability to deploy powered UAVs and parafoils. The GTA would be able to house support materiel, personnel, and MEDVAC modules depending upon mission requirements. This aircraft also could be modified for use as a tanker.

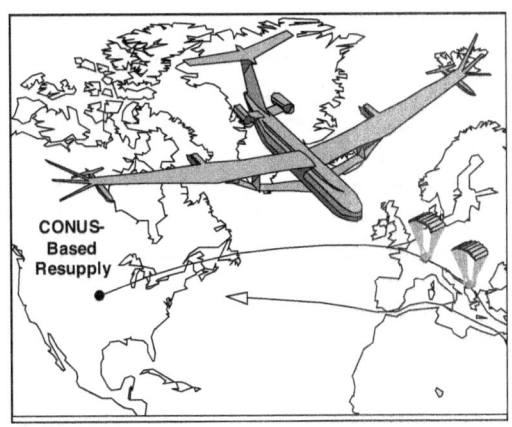

Capabilities

- Payload: 150,000-250,000 lbs
- Unrefueled Range: 10,000-12,000 NM
- Can deploy powered UAVs and parafoils for precision delivery
- Can be reconfigured for air refueling and MEDEVAC

Enabling Technologies (MCTL)

- 1.0, Materials
- 9.1, Gas Turbine Propulsion Systems
- 9.5, Aerospace Structures and Systems
    - 9.5.4, Aircraft High-Performance Structures

White Papers

- F, Airlift

2.1 Strike UAV

Brief Description

The Strike UAV is a low observable, uninhabited air vehicle that loiters subsonically over the region of interest for long periods of time(24+ hours) until directed to strike. Its primary mission is to engage ground targets with standoff precision munitions; however, it also has a limited air-to-air capability. It relies on off-board sensors to supply reconnaissance and targeting information as well as command and control, although it has sufficient on-board sensor capability to allow it to perform preprogrammed missions.

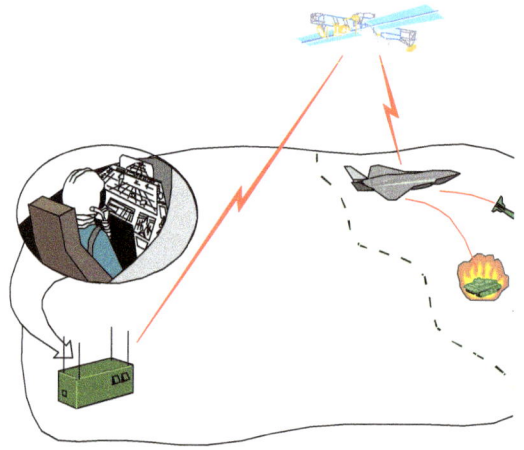

Capabilities

- Subsonic, low observable, long-loiter time (24+ hour) strike capability
- Employs a suite of precision guided munitions
- Air to ground, limited air to air
- Uses primarily off-board sensors
- Capable of operating from in-theater operating bases

Enabling Technologies (MCTL)

- 1.0, Materials
- 4.1, Digital Processing
    - 4.1.4, Image Processing
- 4.2, Software
    - 4.2.5, Data Fusion
- 5.4, Command, Control, Communications, and Intelligence Systems
- 5.5, Information Security
- 9.8, Other Propulsion
- 12.1, Warheads, Ammunition, and Payloads

White Papers

- A, Counter Air
- S, RPV and Aerospace

2.2 Reconnaissance UAV

Brief Description

The uninhabited reconnaissance aerial vehicle (URAV) can be employed either as an independent system or in conjunction with other airborne, ground-based, and space-borne systems. The URAV is fitted with a variety of multispectral sensors, such as infrared, optical, and radar and laser, and collects images, SIGINT, ELINT, and other information. It loiters subsonically at high altitudes over the region of interest for extended periods of time without refueling. The URAV could also be used as part of a bistatic configuration, in which it illuminates the region of interest, while different sensors then receive and process the information.

Capabilities

- High-altitude, long-loiter times (36+ hours)
- Gathers information from multi-spectral sensors, feeds into Global Information Management System

Enabling Technologies (MCTL)

- 1.0 Materials
- 4.1 Digital Processing
    - 4.1.3 Signal Processing
    - 4.1.4 Image Processing
- 4.2 Software
    - 4.2.5 Data Fusion
- 5.1 Transmission
- 6.2 Optical Sensors
- 6.8 Radar
- 6.9 Other Sensors
- 10.3 Power Systems

White Papers

- U, S&R Info Ops
- J, Close Air Support

2.3 Uninhabited Combat Air Vehicle

Brief Description

The uninhabited combat air vehicle (UCAV) can be employed either as an independent system or in conjunction with other airborne, ground-based, and space-based systems. It carries a suite of multispectral sensors (optical, infrared, radar, and laser.) which supplies information to its suite of standoff precision guided munitions. It loiters at high altitude over the region of interest for long periods of time (24+ hours) until called upon to strike a target. While in its subsonic loiter mode, it can perform a surveillance and reconnaissance mission for the global information management system (System 8.1). It could be used as part of a bistatic configuration in which it illuminates a region of interest while a different sensor receives and processes the information. As a secondary mission, it can perform ECM and electronic counter-countermeasures (ECCM) roles.

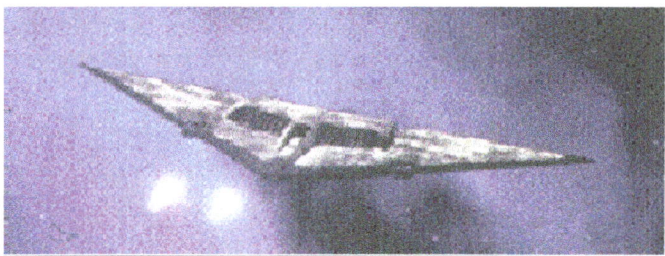

Capabilities

- Subsonic, low observable, long-loiter (24+ hour) reconnaissance and strike capability
- Uses multispectral sensors to employ a suite of precision guided munitions
- Sensors can also feed into the Global Information Management System
- Extensive air-to-ground and air-to-air capability
- Secondary role as ECM and ECCM platform

Enabling Technologies (MCTL)

- 1.0, Materials
- 4.1, Digital Processing
    - 4.1.3, Signal Processing
    - 4.1.4, Image Processing
- 4.2, Software
    - 4.2.5, Data Fusion
    - 4.2.9, Artificial Intelligence
- 5.5, Information Security
- 6.2, Optical Sensors
- 6.8, Radar
- 6.9, Other Sensors
- 9.8, Other Propulsion
- 10.3, Power Systems
- 12.1, Warheads, Ammo, & Payloads

White Papers

- A, Counter Air
- J, Close Air Support
- S, RPV and Aerospace

2.4 Precision Delivery System

Brief Description

A suite of powered and parafoil UAVs capable of autonomous flight for the purpose of all-weather precision airdrop (within 1 meter). High-altitude (40,000 ft) precision airdrops can be achieved using GPS or INS-guided parafoil delivery systems. This technique would allow equipment/supplies to be delivered to forward deployed forces while transport aircraft remain hundreds of miles from the drop zone. Positions can be determined using LIDAR or a GPS instrumented radio drop sound. Powered UAVs would deliver smaller, high-value packages from greater standoff ranges.

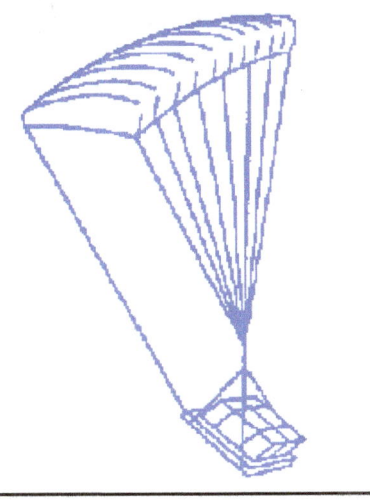

Capabilities

- Precision delivery of equipment or supplies to within one meter accuracy on the ground from 150 NM range
- Uses passive GPS/INS sensors to enhance survivability

Enabling Technologies (MCTL)

- 1.0, Materials Technology
- 2.6, Micromechanical Devices (MEMS)
- 7.3, Vehicle and Flight Control
- 7.1, Inertial Navigation Systems and Related Components
    - 7.1.1, Inertial Navigation Systems
    - 7.1.10, Accelerometers

White Papers

- F, Airlift
- Q, Specail and Humanitarian Operations
- R, Special Operations
- I, Frontier Missions

2.5 UAV Mothership

Brief Description

A large-capacity, long-loiter time, uninhabited subsonic air vehicle used to deploy and recover smaller combat UAVs. It also can replenish them with weapons and propellant. This air vehicle would have the ability to collect, convert and store solar energy and then transfer energy through physical means or by way of beaming to other airborne vehicles, including the Fotofighter (System 1.2).

Capabilities

- A long-loiter air vehicle used to deploy, recover and replenish UAVs
- Uses DEW to serve as a rearming platform for the "Fotofighter"

Enabling Technologies(MCTL)

- 1.0, Materials
- 2.2, Metal Working and Industrial Production
    - 2.2.5, Robots, Controllers, and End-Effectors
- 9.5.4, Aircraft High-Performance Structures
- 10.1, Lasers
- 10.2, Optics
- 10.3, Power systems
- 11.1, High-Energy Laser (HEL) Systems

White Papers

- H, Aerospace Replenishment

2.6 Exfiltration Rocket

Brief Description

The exfiltration rocket (ER) allows a quick extraction of special operations forces teams from the mission area. This system would be brought in during the SOF insertion and assembled at the exfiltration launch site. After mission completion, the SOF team would load themselves and any other items, such as a high-value asset or weapon of mass destruction, into the ER and then take off. The payload and passengers would be recovered by the of an air-retrievable payload system or through a "soft" landing in a friendly area.

Capabilities

- Fast extraction of SOF team from mission area
- 1,000 pounds payload--carries two-four persons, WMD and/or captured high-value asset
- Recovery in air enroute or after soft surface touchdown on friendly ground
- Subsonic speed, 500 NM range, 100 feet CEP

Enabling Technologies (MCTL)

- 1.0, Materials
- 7.3, Vehicle and Flight Control
- 9.5.4, Aerospace Structures and Systems
- 12.7, Energetic Materials

White Papers

- R, Special Operations

3.1 Orbital Maneuvering Vehicle

Brief Description

The orbital maneuvering vehicle (OMV) is an uninhabited orbital propulsion and docking system used to take payloads from an earth-to-orbit lift vehicle and place them in their final orbital plane or used to fetch and return orbiting payloads to a central repair and recovery location. The system also would be capable of carrying line replaceable units (LRU) to a damaged/degraded satellite and accomplishing on-site repair or replacement. It would be designed to allow refueling of civil, commercial, and military satellites as well as the rearming of military space weapons platforms.

Capabilities

- Can maneuver payloads from one earth orbit to another
- Repair of space assets using LPUs
- Refueling of civil, commercial, and military satellites
- Supply/resupply of military space assets

Enabling Technologies (MCTL)

- 2.2, Metal Working and Industrial Production
    - 2.2.5, Robots, Controllers, and End Effectors
- 2.6, Micromechanical Devices (MEMS)
- 4.1, Digital Processing
    - 4.1.1, High-Performance Computing
- 4.2, Software
    - 4.2.9, Artificial Intelligence
- 9.5, Aerospace Structures and Systems
    - 9.5.2, Nonchemical, High-Specific Impulse Propulsion
- 10.3, Power Systems

White Papers

- G, Spacelift
- H, Aerospace Replenishment
- W, On-Orbit Support
- Δ, Space Operations

3.2 Orbital Combat Vehicle

Brief Description

The orbital combat vehicle (OCV) is an uninhabited orbital propulsion and docking system used to take payloads from an earth-to-orbit lift vehicle and place them in their final orbital plane or used to fetch and return orbiting payloads to a central repair and recovery location. The system also would be capable of carrying LRUs to a damaged/degraded satellite and accomplishing on-site repair or replacement. It is designed to allow refueling of civil, commercial, and military satellites as well as the rearming of military space weapons platforms. The OCV is fitted with a medium power high-energy laser system for limited defense and counter-space missions.

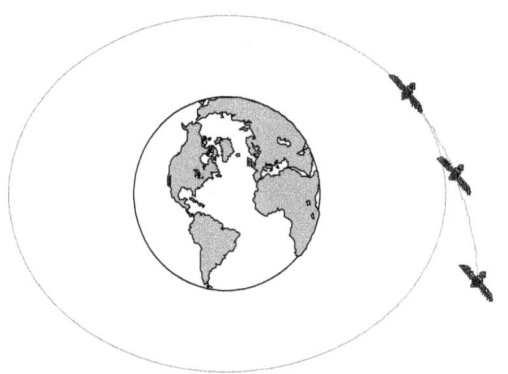

Capabilities

- Refueling of civil, commercial, and military satellites
- Can maneuver payloads from one earth orbit to another
- Supply/resupply of military space assets
- Limited force application against space assets

Enabling Technologies (MCTL)

- 2.2, Metal Working and Industrial Production
    - 2.2.5, Robots, Controllers, and End Effectors
- 2.6, Micromechanical Devices (MEMS)
- 4.1, Digital Processing
    - 4.1.1, High-Performance Computing
- 4.2, Software
    - 4.2.9, Artificial Intelligence
- 9.5, Aerospace Structures and Systems
    - 9.5.2, Nonchemical, High-Specific Impulse Propulsion
- 10.1, Lasers
- 10.3, Power Systems

White Papers

- W, On-Orbit Support
- Δ, Space Operations

3.3 Satellite Bodyguards

Brief Description

A small constellation of defensive satellites (e.g., five) placed in proximity to the protected asset. "Hunter killers" actively seek out threats and incapacitate them with directed energy weapons. Detection of threats from the surface or air would be done by an off-board sensor suite (e.g., systems 8.1 or 8.2) and supplied to the "hunter killer" satellites. Detection of space-based threats would be done by the hunter killer satellites themselves. Decoy satellites appear identical (both EM and visual) to the protected assets to confuse an aggressor; when approached, the decoy can impact and disable the enemy craft.

Capabilities

- Defensive satellites provide protection for high-valued space-based assets
- Active defense hunter killer and decoy satellites seek out and destroy approaching threats

Enabling Technologies (MCTL)

- 4.1, Digital Processing
    - 4.1.1, High-Performance Computing
    - 4.1.3, Signal Processing
- 4.2, Software
    - 4.2.5, Data Fusion
- 9.4, Rockets
- 10.2, Optics
- 10.3, Power Systems
- 11.1, High-Energy Laser Systems

White Papers

- C, Counter space

4.1 Piloted SSTO Transatmospheric Vehicle

Brief Description

This system provides space support and global reach from the earth's surface to low-earth orbit (LEO) using a combination of rocket and hypersonic air-breathing technology. The transatmospheric vehicle (TAV) envisioned takes off vertically, is refuelable in either air or space, and can land on a conventional runway. It has a variable payload capacity (up to 10,000 lbs) and performs as both a sensor and a weapons platform. Alternate missions include satellite deployment and retrieval from LEO and deployment of an ASAT weapon.

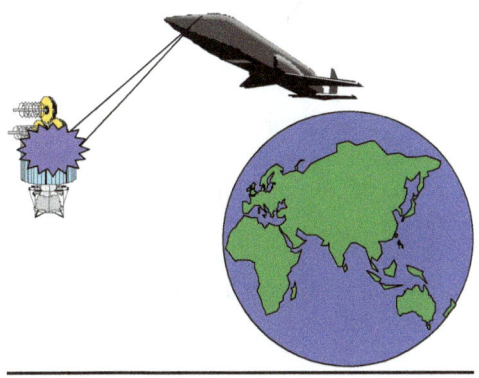

Capabilities

- Single-stage manned TAV capable of providing rapid support to space-based assets from the ground
- Repair or retrieval of satellites in LEO
- Surveillance and reconnaissance platform
- Payload capacity of 10,000lbs
- ASAT weapons platform

Enabling Technologies (MCTL)

- 1.0, Materials
- 4.1, Digital Processing
    - 4.1.1, High-Performance Computing
- 9.2, Ramjet, Scramjet, Combined Cycle Engines
- 9.5, Aerospace Structures and Systems
    - 9.5.4, Aircraft High-Performance Structures
- 12.7, Military Explosives (Energetic Materials)

White Papers

- C, Counter space
- Δ, Space Operations
- F, Airlift
- G, Spacelift
- N, Strategic Attack

4.2 Uninhabited Air-Launched Transatmospheric Vehicle

Brief Description

This uninhabited vehicle is a multirole transatmospheric vehicle (TAV). Launched from an airborne platform (e.g., System 1.1), it is capable of rapid deployment (or retrieval) of satellites providing communication links, and intelligence information. It carries a suite of multi-spectral sensors (optical, infrared, radar, and laser.) for surveillance and reconnaissance missions. This TAV is a rocket-powered vehicle, approximately the size of an F-15 and is capable of carrying several small satellites (e.g., 6 ft by 6 ft by 6 ft, 1,000 lbs each) to low earth orbit (LEO). Further, it could perform an antisatellite (ASAT) role. This TAV can land on a conventional runway.

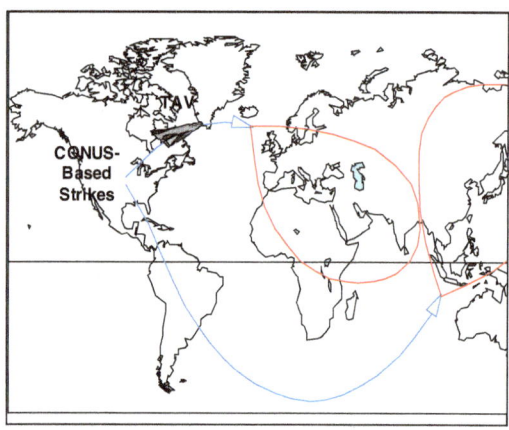

Capabilities

- Rocket-powered "upper-stage" TAV
- Air-launched by hypersonic strike fighter
- Rapid deployment of multiple satellites to LEO
- Repair or retrieval of satellites in LEO
- Surveillance and reconnaissance platform
- Space-based ASAT

Enabling Technologies (MCTL)

- 1.0, Materials
- 4.1, Digital Processing
    - 4.1.1, High-Performance Computing
- 4.2, Software
    - 4.2.5, Data Fusion
- 9.4, Rockets
- 9.5, Aerospace Structures and Systems
    - 9.5.3, Aerospace Thermal Systems
    - 9.5.4, Aircraft High-Performance Structures
- 12.7, Military Explosives (Energetic Materials)

White Papers

- $\Sigma 1$, Space Operations
- Ψ, Hypersonics

5.1 Adjustable Yield Munition

Brief Description

The adjustable yield munition (AYM) is an approach to achieve precise matching of the weapon's effect to the target's characteristics. By manipulating the explosive yield of a weapon (i.e., "dial-a-yield"), collateral damage can be greatly reduced. This is particularly advantageous when flexibility and precision are both required: a platform on patrol, awaiting targets of opportunity, can utilize the same weapon for a hard kill with a large yield, or a surgical, mission-only kill with a tailored yield. One proposed approach to controlling the yield is to change the material composition of the explosive at the molecular level.

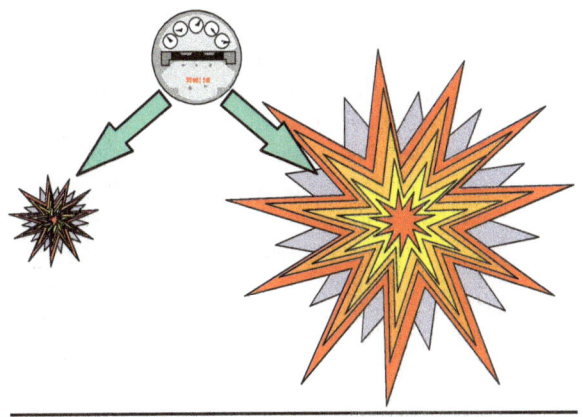

Capabilities

- Precision tailoring of munitions yield to individual target characteristics
- Provides precision strike capability with minimal collateral damage

Enabling Technologies (MCTL)

- 12.7, Military Explosives (Energetic Materials)

White Papers

- K, Interdiction

5.2 Advanced Air-to-Air Missile

Brief Description

This long-range air-to-air missile receives real-time target information from off-board sensors and utilizes reactive jets and an on-board computer to acquire, pursue, and destroy enemy air assets including cruise missiles. Terminal tracking and guidance may employ a combination of LIDAR, infrared (IR), radio frequency (RF), magnetic anomaly detection (MAD), jet engine modulation (JEM), photographic and acoustic sensors.

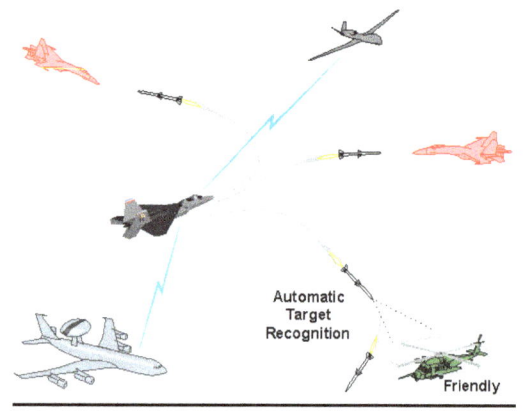

Capabilities

- 100+ NM range air-to-air missile that can engage and destroy airborne targets with a high probability of kill (> .9)
- Counters low observable (LO) targets
- Counter countermeasures enable this system to target high-value enemy surveillance and control and jamming airborne platforms

Enabling Technologies (MCTL)

- 2.6, Micromechanical Devices
- 4.1, Digital Processing
    - 4.1.1, High-Performance Computing
- 4.2, Software
    - 4.2.5, Data Fusion
- 6.1, Air, Marine, Space Platform and Terrestrial Acoustic Systems
- 6.2, Optical Sensors
- 6.4, Electronic Combat (EC)
- 6.5, Magnetometers and Magnetic Gradiometers
- 6.8, Radar
- 6.9, Other Sensors
- 7.3, Vehicle and Flight Control

White Papers

- A, Counter air

5.3 Airborne High-Power Microwave Weapon

Brief Description

A pulsed power airborne high-power microwave (HPM) system is proposed. This medium range weapons system would constitute the primary payload of the host escort defense aircraft. The system generates variable magnitude HPM fields that disrupt or destroy electrical components in the target region. The HPM weapon envisioned is capable of engaging both air and ground targets.

Capabilities

- Aircraft defense against weapon systems and munitions having sensitive electrical components.
- The ability to strike against air and surface targets with variable lethality.

Enabling Technologies (MCTL)

- 1.0, Materials
- 10.3, Power Systems
- 11.2, High-Power Radio Frequency (RF) Systems

White Papers

- K, Interdiction
- A, Counterair

5.4 Standoff Hypersonic Missile

Brief Description

This hypersonic air-to-ground missile is launched from a hypersonic strike vehicle (System 1.1) and utilizes a Scramjet to propel itself at Mach 8 toward the intended high-value target. It then glides to target at Mach 4 and its flight trajectory altered as needed by way of off-board control. Its high speed air-launched range is 1,450 NM.

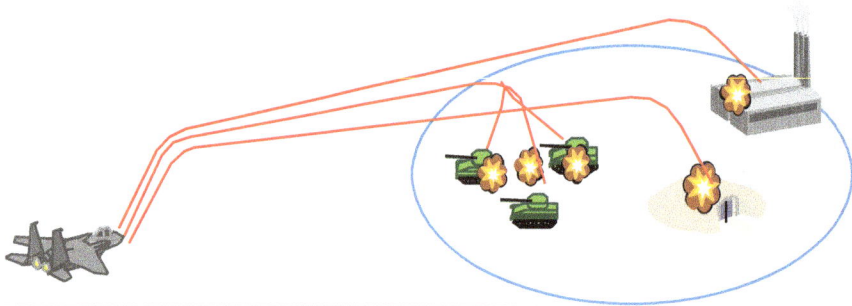

Capabilities

- Hypersonic theater-range precision ground strike munition
- Launched at Mach 8, its range is 1450 NM
- Terminally guided to the target by way of off-board control

Enabling Technologies (MCTL)

- 1.2, Composite Materials
- 5.1, Transmission
- 9.2, Ramjet, Scramjet, and Combined Cycle Engines
- 9.5.4, Aircraft High-Performance Structures
- 12.7, Energetic Materials

White Papers

- Ψ, Hypersonics

5.5 Attack Microbots

Brief Description

"Attack microbots" describes a class of highly miniaturized (one millimeter scale) electromechanical systems capable of being deployed en masse and performing individual or collective target attack. Various deployment approaches are possible, including dispersal as an aerosol, transportation by a larger platform, and full flying/crawling autonomy. Attack is accomplished by a variety of robotic effectors, electromagnetic measures, or energetic materials. Some "sensor microbot" capabilities are required for target acquisition and analysis.

Capabilities

- "Swarm" of 1mm scale flight-capable MEM platforms provide unobtrusive, pervasive intervention into adversary environments and systems
- Extremely small size provides high-penetration capabilities and natural stealth

Enabling Technologies (MCTL)

- 2.2.5, Robots, Controllers and End-Effectors
- 2.6, Micromechanical Devices (MEMS)
- 4.1, Digital Processing
    - 4.1.1, High-Performance Computing
- 4.2, Software
    - 4.2.5, Data Fusion
- 5.1, Transmission
- 5.4, C^4I
- 7.3, Vehicle and Flight Control
- 10.3, Power Systems
- 12.7, Energetic Materials

White Papers

- A, Counterair
- C, Counterspace
- Q, Special and Humanitarian Operations

5.6 Airborne Holographic Projector

Brief Description

The holographic projector displays a three-dimensional visual image in a desired location, removed from the display generator. The projector can be used for psychological operations and strategic perception management. It is also useful for optical deception and cloaking, providing a momentary distraction when engaging an unsophisticated adversary.

Capabilities

- Precision projection of 3-D visual images into a selected area
- Supports PSYOP and strategic deception management
- Provides deception and cloaking against optical sensors

Enabling Technologies (MCTL)

- 4.1.4, Image Processing (holographic displays)
- 10.1, Lasers
- 10.2, Optics
- 10.3, Power Systems

White Papers

- Q, Special & Humanitarian Operations
- N, Strategic Attack

5.7 Hybrid High-Energy Laser System

Brief Description

The hybrid high-energy laser system (HHELS) consists of several ground-based, multi-megawatt high-energy chemical lasers and a constellation of space-based mirrors. The HHELS can be used in several modes of operation. In its weapons mode with the laser at high power, it engages air, space and ground targets by reflecting a laser beam off one or more of the mirrors to the intended target. It also can be used for target tracking, limited space debris removal(1-10 centimeter objects), and replenishment of satellites.

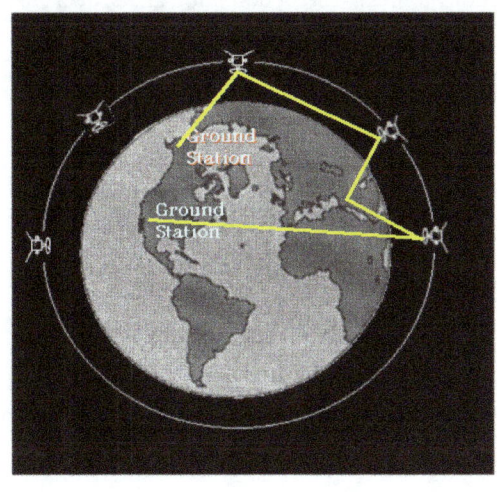

Capabilities

- Worldwide coverage provided by several ground laser sites and a constellation of 15-20 space-based mirrors
- Has counterspace, counterair, force application, and weather modifica-tion uses
- Replenishes some space-based assets
- Ground laser sites have a limited space surveillance capability

Enabling Technologies (MCTL)

- 9.5, Aerospace Structures and Systems
    - 9.5.1, Spacecraft Structures
    - 9.5.2, Nonchemical, High-Specific Impulse Propulsion
- 10.2, Optics
- 10.3, Power Systems
- 11.1, High-Energy Laser Systems

White Papers

- Δ, Space Operations

6.1 Global Area Strike System

Brief Description

The global area strike system (GLASS) consists of a high-energy laser (HEL) system, a kinetic energy weapon (KEW) system and a transatmospheric vehicle (TAV). The HEL system consists of ground-based lasers and space-based mirrors which direct energy to the intended target. The KEW system (System 6.2) consists of terminally guided projectiles, with and without explosive enhancers. The TAV (System 4.1) is a flexible platform capable of supporting maintenance and replenishment of the HEL and KEW space assets and also could be used for rapid deployment of special operations forces. Target definition and sequencing is managed externally (i.e., using GIMS (System 8.1)).

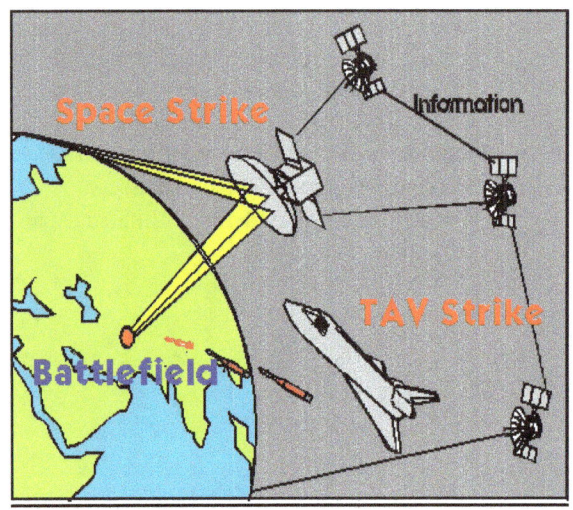

Capabilities

- Uses a constellation of space-based KEWs and hybrid (ground-based) high-energy lasers to provide precision global engagement of ground, space and airborne targets, with variable lethality
- Provides extensive surveillance capability using TAVs and ground-based laser sites
- Rapid deployment from ground to LEO using TAV fleet
- Limited weather modification uses

Enabling Technologies (MCTL)

- 1.0, Materials
- 2.6, Micromechanical Devices
- 4.1, Digital Processing
    - 4.1.1, High-Performance Computing
- 5.1, Transmission
- 7.3, Vehicle and Flight Control
- 9.2, Ramjet, Scramjet, Combined Cycle Engines
- 9.5, Aerospace Structures & Systems
    - 9.5.1, Spacecraft Structures
    - 9.5.4, Aircraft High-Performance Structures
- 10.2, Optics
- 11.1, High-Energy Laser Systems
- 11.4, Kinetic Energy Systems
    - 11.4.2, Kinetic Energy Projectiles
    - 11.4.4, Kinetic Energy Platform Management
- 12.7, Energetic Materials

White Papers

- Δ, Space Operations

6.2 Space-Based Kinetic Energy Weapon

Brief Description

The space-based KEW is a general class of LEO based weapons that include a variety of warhead types from flechettes and pellets to large and small high-density rods. The KEW may be directed at air, space, and ground targets and achieves its destructive effect by penetrating the target at hypervelocity. Sensor information is provided to the KEW by a main sensor suite off-board of the vehicle (such as GSRT [System 8.2] or GIMS [System 8.1]). However, each armament has a minimal sensor capability (i.e., GPS receiver) and a simple flight control system for maneuver.

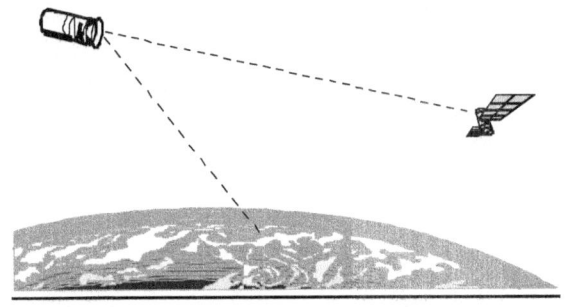

Capabilities

- Precision global engagement of space, air, and ground targets using a constellation of space-based KEW platforms
- Micro and larger space-delivered kinetic energy munitions
- Provide concentrated or dispersed target coverage

Enabling Technologies (MCTL)

- 1.0, Materials
- 2.6, Micromechanical Devices (MEMS)
- 4.1, Digital Processing
  - 4.1.1, High-Performance Computing
- 5.1, Transmission (Communications)
- 7.3, Vehicle and Flight Control
- 11.4, Kinetic Energy Systems
  - 11.4.2, Kinetic Energy Projectiles
  - 11.4.4, Kinetic Energy Platform Management

White Papers

- C, Counterspace
- N, Strategic Attack
- Δ, Space Operations

6.3 Space-Based High-Power Microwave Weapon

Brief Description

The space-based high-power microwave (HPM) weapon system is capable of engaging ground, air and space targets with a varying degree of lethality. It consists of a constellation of satellites deployed in LEO (approx 500 NM) that can direct an ultra wide-band (UWB) of microwave energy at ground, air, and space targets. Its effect is to generate high-electric fields over a target area tens to hundreds of meters in size, thereby disrupting or destroying any electronic components present.

Capabilities

- Engage ground, space and airborne targets containing sensitive electrical components
- Variable lethality, from disrupt to destroy
- Provides limited weather modifica-tion capability

Enabling Technologies (MCTL)

- 1.0, Materials
- 4.1, Digital Processing
    - 4.1.1, High-Performance Computing
- 9.5, Aerospace Structures and Systems
    - 9.5.1, Spacecraft Structures
- 10.3, Power Systems
- 11.2, High-Power Radio Frequency Systems

White Papers

- A, Counterair
- C, Counterspace
- Q, Special and Humanitarian Operations
- Γ, Weather Modification
- Δ, Space Operations

6.4 Space-Based High-Energy Laser System

Brief Description

The space-based high energy laser system is a space-based, multimegawatt, high-energy chemical laser constellation that can be used in several modes of operation. In its weapons mode with the laser at high power, it can attack ground, air, and space targets. In its surveillance mode, it can operate using the laser at low-power levels for active illumination imaging or with the laser inoperative for passive imaging.

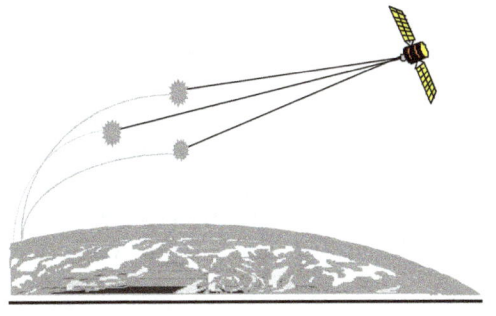

Capabilities

- Worldwide coverage provided by constellation of 15-20 chemical HELs
- Provides optical surveillance of air, space, and ground objects using active or passive imaging
- Provides precision strike of air, space, and ground targets with variable lethality
- Provides limited weather modifica-tion capability

Enabling Technologies (MCTL)

- 4.1, Digital Processing
    - 4.1.1, High-Performance Computing
    - 4.1.3, Signal Processing
    - 4.1.4, Image Processing
- 4.2, Software
    - 4.2.4, Hard Real-Time Systems
    - 4.2.5, Data Fusion
- 10.2, Optics
- 10.3, Power Systems
- 11.1, High-Energy Laser Systems

White Papers

- A, Counterair
- C, Counterspace

6.5 Solar-Powered High-Energy Laser System

Brief Description

The solar-powered high-energy laser system is a space-based, multimegawatt, high energy solar-powered laser constellation that can be used in several modes of operation. In its weapons mode with the laser at high power, it can attack ground, air, and space targets. In its surveillance mode, it can operate using the laser at low-power levels for active illumination imaging, or with the laser inoperative for passive imaging.

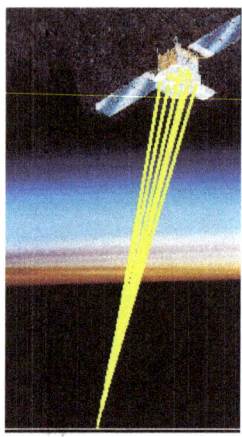

Capabilities

- Variable lethality, space-based solar-powered high-energy laser constel-lation
- Provides global optical surveillance by active or passive imaging
- Engages ground, space, and airborne targets
- Possible application in weather modification

Enabling Technologies (MCTL)

- 4.1, Digital Processing
    - 4.1.1, High-Performance Computing
    - 4.1.3, Signal Processing
    - 4.1.4, Image Processing
- 4.2, Software
    - 4.2.4, Hard Real-Time Systems
    - 4.2.5, Data Fusion
- 10.2, Optics
- 10.3, Power Systems
- 11.1, High-Energy Laser Systems

White Papers

- C, Counterspace
- Δ, Space Operations

6.6 Solar Energy Optical Weapon

Brief Description

The solar energy optical weapon (SEOW) consists of a constellation of space-based mirrors which allow solar radiation to be focused on specific ground, air, or space targets. The lethality of this system is limited, due to optical diffusion; however, it may prove useful for disruption or perhaps weather control.

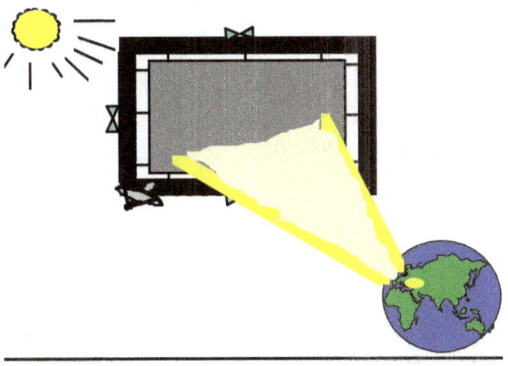

Capabilities

- Space-based solar "flashlight"
- Illuminates air, ground, and space targets with focused solar radiation
- Engage space-based targets with variable lethality (heating caused by focused solar radiation)
- Focuses solar energy on ground-based collectors
- Could potentially be used for weather modification

Enabling Technologies (MCTL)

- 2.2, Metal Working and Industrial Production
    - 2.2.5, Robots, Controllers, and End Effectors
- 9.5, Aerospace Structures and Systems
    - 9.5.1, Spacecraft Structures
- 10.2, Optics
- 10.3, Power Systems

White Papers

- C, Counterspace
- Δ, Space Operations

6.7 Asteroid Mitigation System

Brief Description

The asteroid mitigation system protects the Earth/Moon system from earth-crossing objects (ECO) by either deflecting or fragmenting ECO they no longer pose a threat. Deflection could be accomplished using nuclear explosive devices.

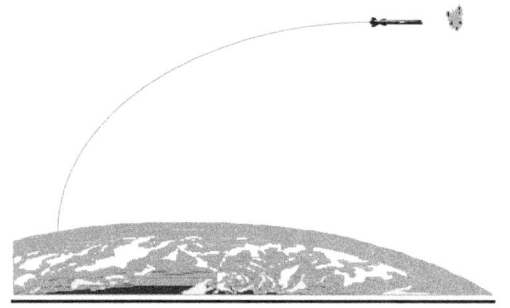

Capabilities

- Deflects or destroys objects in space having the size and trajectory to threaten the Earth/Moon system

Enabling Technologies (MCTL)

- 7.3, Vehicle and Flight Control
- 9.5, Aerospace Structures and Systems
    - 9.5.2, Nonchemical, High "Specific Impulse" Propulsion
- 12.7, Military Explosives (Energetic Materials)

White Papers

- B, Planetary Defense

7.1 Spoken Language Translator

Brief Description

This hand-held or worn device translates oral communications in near real time. It would enhance multinational operational effectiveness in all areas, including training, diplomacy, special operations, and conventional ground operations. It is capable of one-for-one word substitution in a wide variety of languages and provides two-way communications between the owner and another person. The system will have a limited ability to compensate for differences in sentence syntactic structures, cultures, dialects, and idioms/slang and a limited ability to select words according to context. Careful placement of both microphones and both speakers is required for deconfliction (not having to hear both languages simultaneously), limiting the scope of its operation; the system is best suited for much controlled two-way communications as telephone, radio, or computer. The system also would be useful for written text translation.

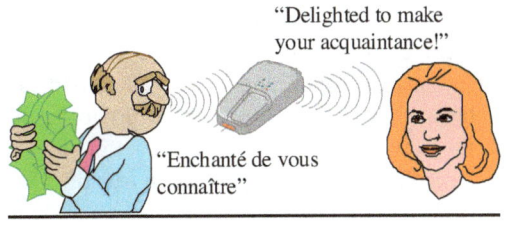

Capabilities

- Near real time speech translation
- Bi-directional, multilingual

Enabling Technologies (MCTL)

- 4.1, Digital Processing
    - 4.1.1, High-Performance Computing
    - 4.1.6, Speech-Processing Systems
- 4.2, Software
    - 4.2.9, Artificial Intelligence

White Papers

- Q, Special and Humanitarian Operations

7.2 Personal Digital Assistant

Brief Description

The personal digital assistant (PDA) connects an individual to the information systems of 2025. PDA is envisioned as a hand held or wrist-watch size unit. Possible input modes include both touch and voice. It is the warrior's secure, high-capacity connection to the distributed C^4I system. The PDA maintains the owner's personal data such as medical and training records. It learns and remembers the owner's preferences and needs so that requests for information are properly tailored. It is self-securing: it recognizes the owner through a number of biometrics and ensures that it cannot be commandeered. In short, the PDA is a single device that can replace the cellular telephone, radio, personal computer, identification and banking cards, and any other personal information-management device of today.

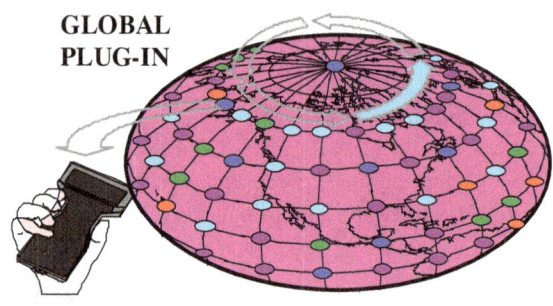

GLOBAL PLUG-IN

Capabilities

- Warriors can attain situational awareness, act as a sensor (reconnaissance) node, and receive instructions/ commands.
- Commanders can monitor the status of their troops by woay of the physiological measures relayed by the PDA.
- It also contains an identification friend or foe (IFF) module.

Enabling Technologies (MCTL)

- 2.6, Micromechanical Devices
- 4.1, Digital Processing
    - 4.1.1, High-Performance Computing
    - 4.1.3, Signal Processing
    - 4.1.6, Speech Processing
- 4.2, Software
    - 4.2.9, Artificial Intelligence
- 5.1, Transmission
- 5.3, Communications Network Management and Control
- 5.5, Information Security
- 10.3, Power Systems

White Papers

- J, Close Air Support
- K, Interdiction
- M, Information Operations
- Q, Special and Humanitarian Operations
- R, Special Operations
- T, Space S&R Fusion
- U, S&R Information Operations
- V, S&R Real Time Integration
- Z, Information Technology Integration in Education and Training
- Θ, Combat Support

7.3 Virtual Interaction Center

Brief Description

The Virtual Interaction Center describes a virtual reality environment in which commanders can immerse themselves in a three-dimensional representation of the battlespace. Information from a global information system, including GIMS (System 8.1), is displayed in a virtual reality environment, giving the commander situational awareness. The center also has the capability to replay battles and engagements and to simulate "what if" scenarios.

Capabilities

- Immersive virtual reality environ-ment to view and control battlespace
- Provides commanders with situa-tional awareness and C^4 functions in a collaborative environment
- Simulation capability for "what if" exercises
- Simulation and virtual reality environment for education and training

Enabling Technologies (MCTL)

- 4.1, Digital Processing
    - 4.1.1, High-Performance Computing
    - 4.1.2, Dynamic Training and Simulation
    - 4.1.4, Image Processing
- 4.2, Software
    - 4.2.4, Hard Real-Time Systems

White Papers

- A, Counterair
- J, Close Air Support
- K, Interdiction
- M, Information Operations
- N, Strategic Attack
- Q, Special and Humanitarian Ops
- U, S&R Information Operations
- V, S&R Real-Time Integration
- W, On-Orbit Support
- X, General Education & Training
- Z, Information Technology Integra-tion in Education & Training
- Θ, Combat Support
- Σ2, AFIT Space Operations

8.1 Global Information Management System

Brief Description

The global information management system is a pervasive network of intelligent information-gathering, processing, analysis, and advisory nodes. It collects, stores, analyzes, fuses, and manages information from ground/air/space sensors and all source intelligence. All types of sensors (i.e., acoustic, optical, radio frequency, and olfactory) are used. However, the true power of this system is its use of neural processing to provide the right type of information based on the user's personal requirements.

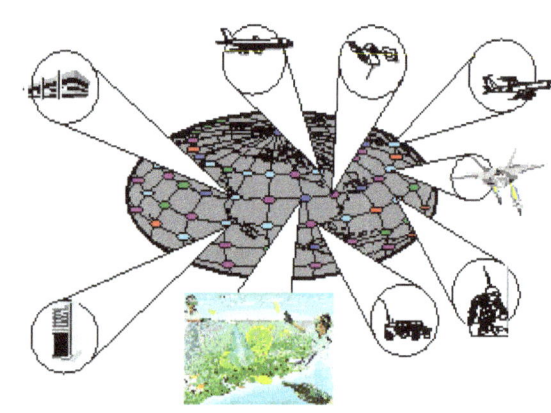

Capabilities

- Provides situational and battle-space awareness, tailored to each user's needs and interests
- Provides extensive information warfare capability
- Uses neural networks, artificial intelligence, and intelligent software agents to gather, synthesize, and format information
- Provides human interfaces through Personal Digital Assistants, Virtual Interaction Centers, and other systems

Enabling Technologies (MCTL)

- 2.6, Micromechanical Devices
- 4.1, Digital Processing
  - 4.1.1, High-Performance Computing
  - 4.1.2, Dynamic, Training, and Simulation
  - 4.1.3, Signal Processing
  - 4.1.4, Image Processing
- 4.2, Software
  - 4.2.4, Hard Real-Time Systems
  - 4.2.5, Data Fusion
  - 4.2.9, Artificial Intelligence
- 5.1, Transmission
- 5.3, Communications Network Management and Control
- 5.4, Command, Control, Communications, and Intelligence Systems
- 5.5, Information Security
- 5.3, Communications Network Management and Control
- 6.1, Air, Marine, Space Platform, and Terrestrial Acoustic Systems
- 6.2, Optical Sensors
- 6.6, Magnetometers and Magnetic Gradiometers
- 6.7, Gravity Meters and Gravity Gradiometers
- 6.8, Radar
- 6.9, Other Sensors

White Papers

- A, Counterair
- B, Planetary Defense
- C, Counterspace
- J, Close Air Support
- K, Interdiction
- M, Information Operations
- N, Strategic Attack
- T, Space S&R Fusion
- U, S&R Information Operations
- V, S&R Real-Time Integration
- W, On-Orbit Support
- Θ, Combat Support
- Σ2, AFIT Space Operations
- Φ, Logistics

8.2 Global Surveillance, Reconnaissance, and Targeting System

Brief Description

The global surveillance, reconnaissance, and targeting system (GSRT) is a space-based omnisensorial collection, processing, and dissemination system to provide a real-time information database. This database is used to create a virtual reality image of the area of interest. This image can be used at all levels of command to provide situational awareness, technical and intelligence information, and two-way command and control.

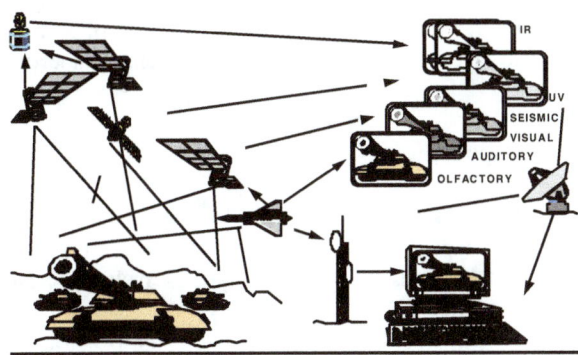

Capabilities

- Real-time information on demand to war fighter
- Smaller, distributed, proliferated satellites and space-based sensors
- Omnisensorial collection, processing and information dissemination
- Space-based fusion and data storage

Enabling Technologies (MCTL)

- 2.6, Micromechanical Devices (MEMS)
- 4.1, Digital Processing
    - 4.1.1, High-Performance Computing
    - 4.1.3, Signal Processing
    - 4.1.4, Image Processing
- 4.2, Software
    - 4.2.4, Hard Real-Time Systems
    - 4.2.5, Data Fusion
- 5.1, Transmission
- 5.3, Communications Network Management and Control
- 5.4, Command, Control, Communications, and Intelligence Systems
- 5.5, Information Security
- 6.1, Air, Marine, Space Platform, and Terrestrial Acoustic Systems
- 6.2, Optical Sensors
- 6.6, Gravity Meters and Gravity Gradiometers
- 6.7, Radar

White Papers

- W, On-Orbit Support

8.3 Sensor Microbots

Brief Description

Sensor microbots describes a class of highly miniaturized (millimeter-sized) electromechanical air and ground systems capable of being deployed en masse to collect data, perform individual and collective data fusion, and communicate that data for further processing and distribution. Various deployment approaches are possible, including dispersal as an aerosol, transportation by a larger platform, and full-flying/crawling autonomy. Data collection is accomplished through miniaturized onboard sensors, typically restricted to one or two sensors per unit due to size and power limitations. Communications are possible by transmission through relay stations "relaybots" or physical collection of the microbots. Some applications of sensor microbots are: security net to guard own assets, surveillance and reconnaissance, and intelligence gathering on adversary assets.

Capabilities

- "Swarm" of one millimeter scale, flight-capable MEM platforms provide unobtrusive, pervasive, multi-spectral sensing
- Small size provides high-penetration capabilities and natural stealth

Enabling Technologies (MCTL)

- 2.2, Metal Working and Industrial Production
    - 2.2.5, Robots, Controllers and End Effectors
- 2.6, Micromechanical devices (MEMS)
- 4.1, Digital Processing
    - 4.1.1, High Performance Computing
    - 4.1.3, Signal Processing
    - 4.1.4, Image Processing
- 4.2, Software
    - 4.2.5, Data Fusion
    - 4.2.9, Artificial Intelligence
- 5.1, Transmission
- 5.4, Command, Control, Communications, and Intelligence Systems
- 6.1, Acoustic Sensors
- 6.2, Optical Sensors
- 6.4, Electronic Combat
- 6.8, Radar

- 6.9, Other Sensors
- 7.3, Vehicle and Flight Control
- 9.5, Aerospace Structures and Systems
- 10.3, Power Systems

White Papers

- A, Counterair
- D, Operability and Defense
- J, Close Air Support
- K, Interdiction
- M, Information Operations
- N, Strategic Attack
- Q, Special and Humanitarian Operations
- T, Space S&R Fusion
- U, S&R Information Operations
- V, S&R Real-Time Integration
- Θ, Combat Support

8.4 Multiband Laser Sensor System

Brief Description

Different frequencies of electromagnetic energy vary in their ability to penetrate materials. For a particular material, one frequency will reflect off the surface, another will penetrate. By employing a suite of laser devices over a wide frequency range, complete internal and external inspection of a structure can be accomplished and a full three-dimensional model can be developed. This tool can be used for nondestructive inspection of components, target vulnerability analysis, target identification and decoy rejection, and reconnaissance. As envisioned, this suite of laser devices would be carried on an airborne platform, but it clearly has ground-based applications also.

Capabilities

- Airborne variable-frequency laser pod provides internal examination of materials and structures
- Precision target vulnerability analysis
- Target identification and decoy rejection
- Structural mapping for reconnais-sance
- Nondestructive testing of parts for "right-timed" preventative mainte-nance

Enabling Technologies (MCTL)

- 4.1, Digital Processing
    - 4.1.1, High-Performance Computing
    - 4.1.4, Image Processing
- 10.1, Lasers
- 10.2, Optics
- 10.3, Power Systems

White Papers

- K, Interdiction

8.5 Asteroid Detection System

Brief Description

The asteroid detection system is a network of ground and space sensors which search for, track, and characterize space objects which are large enough and in an orbit to threaten the earth-moon system. The system also includes a centralized processing center which fuses data from all of the available sensors, catalogs the known objects, and distributes information to the known authorities.

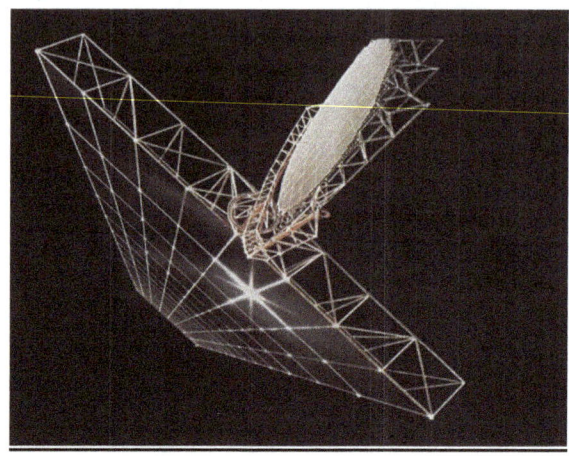

Capabilities

- Deep space surveillance – the ability to locate 100-meter-diameter objects at a minimum of one astronomical unit from earth using radar, optical/infrared sensor platforms
- Track and maintain historical data on asteroids and comets
- Provide near real-time feedback on the effects of asteroid mitigation systems (similar to real time BDA)

Enabling Technologies (MCTL)

- 2.6, Micromechanical Devices
- 4.1, Digital Processing
    - 4.1.1, High-Performance Computing
    - 4.1.4, Image Processing
- 4.2, Software
    - 4.2.5, Data Fusion
- 9.5, Aerospace Structures & Systems
    - Spacecraft Structures

White Papers

- B, Planetary Defense

9.1 Mobile Asset Repair Station

Brief Description

In wartime, replacement parts will be repaired or manufactured in the theater of operations for a variety of deployed weapon systems through the mobile asset repair station (MARS). MARS is a concept whereby parts can be repaired or manufactured using a mobile facility which can be land-based or water-based in the theater of operations, but out of harm's way. The facility features a set of fully integrated flexible manufacturing systems (FMS) and robotic systems that are linked to the commercial manufacturers. These manufacturers supply the specifications to the FMS which then produces the part or component. Many of the required materials necessary for MARS to manufacture the components will be obtained from local countries.

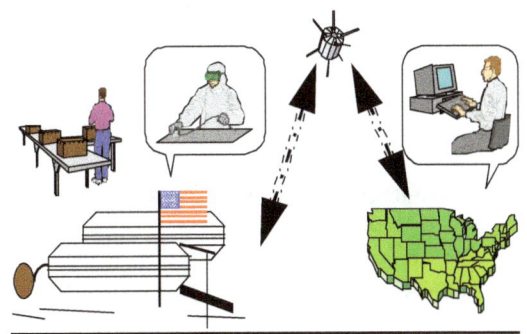

Capabilities

- Manufacture avionics and mechanical system components in-theater
- Uses flexible manufacturing and robotic systems electronically linked to commercial manufacturers
- Cut, repair, and manufacture turn-around times significantly by eliminating need to transport failed (or replacement) parts back to (or from) depot or commercial supplier

Enabling Technologies (MCTL)

- 1.0, Materials
- 2.1, Automation of Industrial Processes, Systems, and Factories
    - 2.1.1, Computer-Aided Design/Engineering and Interface with Computer- Aided Manufacturing
    - 2.1.2, Computer-Aided Manufacturing, Inspection, and Testing
    - 2.1.3, Computer-Aided Servicing and Maintenance
- 2.2, Metal Working and Industrial Production
    - 2.2.1, Numerically Controlled Machine Tools
    - 2.2.5, Robots, Controllers, and End Effectors
- 2.6, Micromechanical Devices (MEMS)

White Papers

- Ω, Logistics

9.2 Weather Analysis and Modification System

Brief Description

A global network of sensors provides "weather warriors" with the means to monitor and accurately predict weather activities and their effects on military operations. A diverse set of weather modification tools allows manipulation of small-to-medium scale weather phenomena to enhance friendly force capabilities and degrade those of the adversary. Many of the sensors required for this system are assumed to be external (i.e., part of the global information management system (GIMS), discussed in System 8.1).

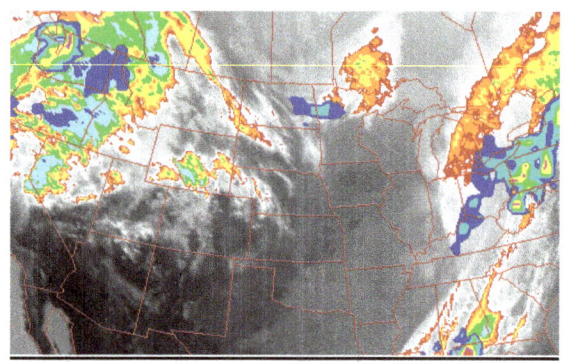

Capabilities

- Understanding and predicting local weather effects on military oper-ations
- Precipitation inducement or suppres-sion using particulate seeding or directed energy
- Fog generation/dissipation using directed energy techniques
- Storm triggering/enhancement using airborne cloud seeding
- High-power microwave (HPM) devices (ground-based) and iono-spheric mirrors for communications and radar enhancement/disruption
- Ionospheric charging for spacecraft disruption using crossed HPM beams

Enabling Technologies (MCTL)

- 2.6, Micromechanical Devices
- 4.1, Digital Processing
    - 4.1.1, High-Performance Computing
- 4.2, Software
    - 4.2.5, Data Fusion
    - 4.2.9, Artificial Intelligence
- 4.3, Hybrid Computing
- 6.9 Other Sensors
- 11.2, High-Power RF Systems

White Papers

- Γ, Weather Modification

9.3 Sanctuary Base

Brief Description

The Sanctuary Base provides a secure, low observable, all-weather forward operating base that reduces the number of assets requiring protection from attack. The runway, power systems, ordnance storage, aircraft maintenance assets, and C^4I systems are self-maintaining and self-repairing. Base security is highly automated. Chemical/biological hazards are cleaned up by nanobots and biotechnology. Robots perform refueling, weapons-loading, maintenance, security, and explosive ordnance destruction.

Capabilities

- Self-defending, self-repairing, and secure
- Facilities, taxiways, and runways are self-diagnostic and self-healing
- Sensors and weapons suite provide complete security within a 500 NM radius
- Orbiting UAVs, robots, and force projectors (DEW/kinetic) provide active defense
- Robots perform refueling, weapons loading, maintenance, EOD, and decontamination
- Low observable buildings and facilities
- Contains command center using the Virtual Interaction Center

Enabling Technologies (MCTL)

- 1.0, Materials
- 2.1, Automation of Industrial Processes, Systems, and Factories
- 2.2, Metal Working and Industrial Processes, Systems, and Factories
    - 2.2.5, Robots, Controllers, and End Effectors
- 2.6, Micromechanical Devices
- 4.1, Digital Processing
    - 4.1.1, High-Performance Computing
    - 4.1.2, Dynamic Training and Simulation
    - 4.1.4, Image Processing
- 4.2, Software
    - 4.2.4, Hard Real-Time Systems
    - 4.2.9, Artificial Intelligence
- 10.1, Lasers
- 10.3, Power Systems
- 11.1, High-Energy Laser (HEL) Systems
- 11.2, High-Power Radio Frequency (RF) Systems
- 11.4, Kinetic Energy (KE) Systems
- 11.4.4, Kinetic Energy Platform Management
- 13.3, Chemical/Biological Warfare Defensive Systems

White Papers

- D, Operability and Defense

Appendix C

Alternate Futures Weights

This appendix presents the value model weights for each of the six alternate futures. There are two sets of weights for each future. One set is the average of the weights given by the student members of the *AF 2025* writing teams and is denoted "AU Team Weights" (fig. C-1 through fig. C-6). The other set was developed by the *AF 2025* Alternate Futures team and is denoted "Alt Futures Team Weights" (fig. C-7 through fig. C-12) The final three figures (fig. C-13 through fig. C-15) contain the weights for the value model force qualities. These were the same for both weight sets across all alternate futures.

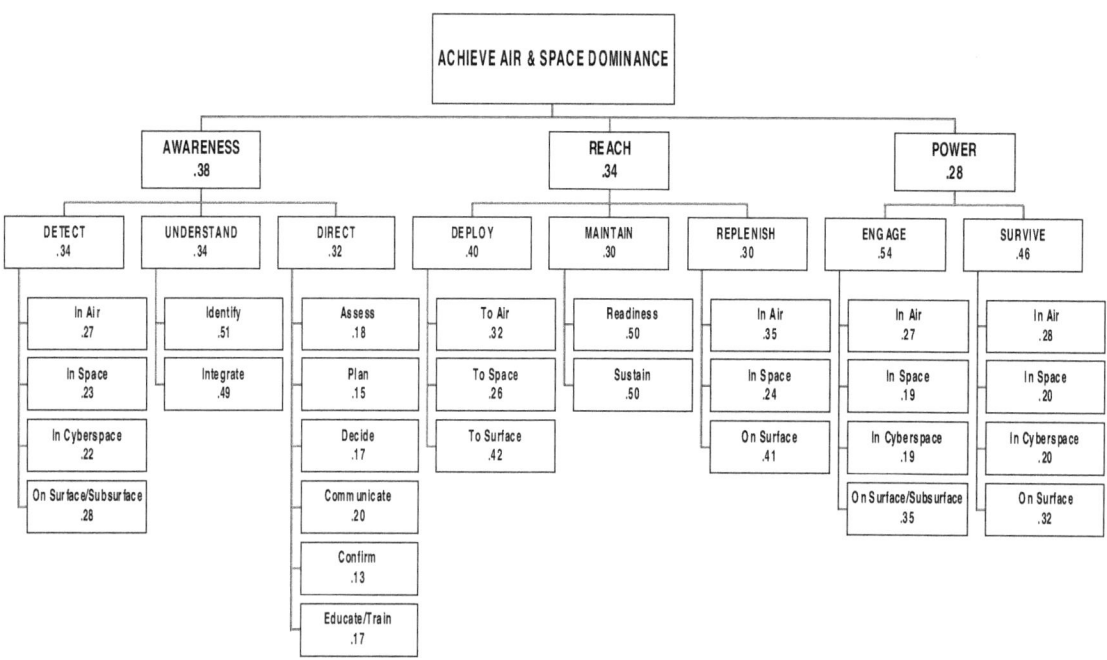

Figure C-1. AU Team Weights - Halfs Future

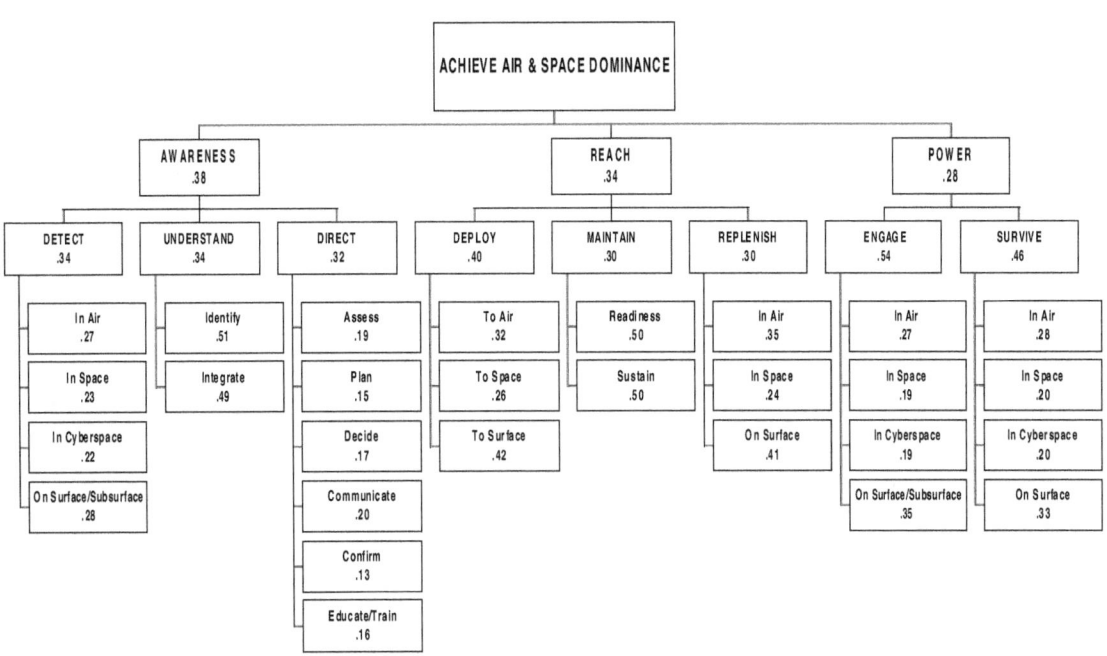

Figure C-2. AU Team Weights - Gulliver's Future

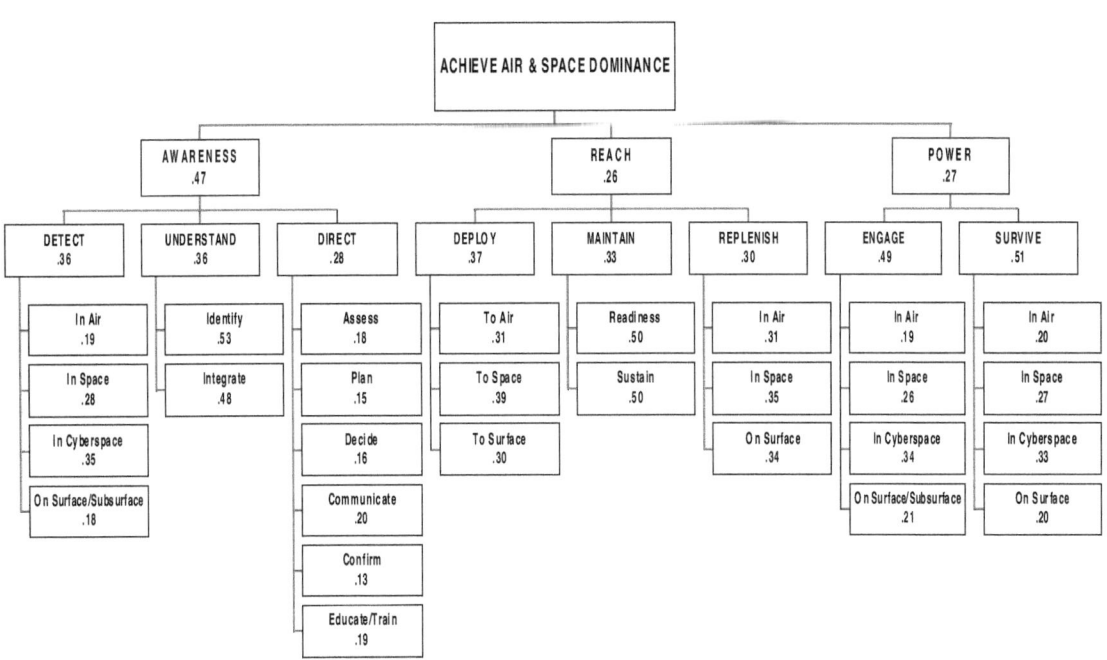

Figure C-3. AU Team Weights - Zaibatsu Future

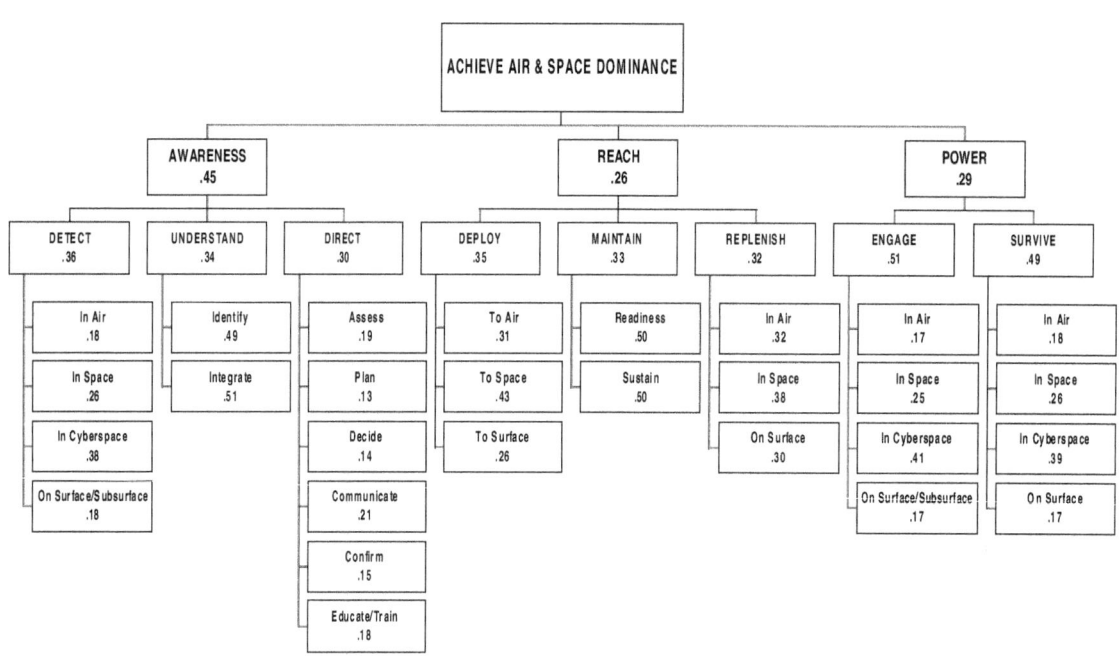

Figure C-4. AU Team Weights - Digital Future

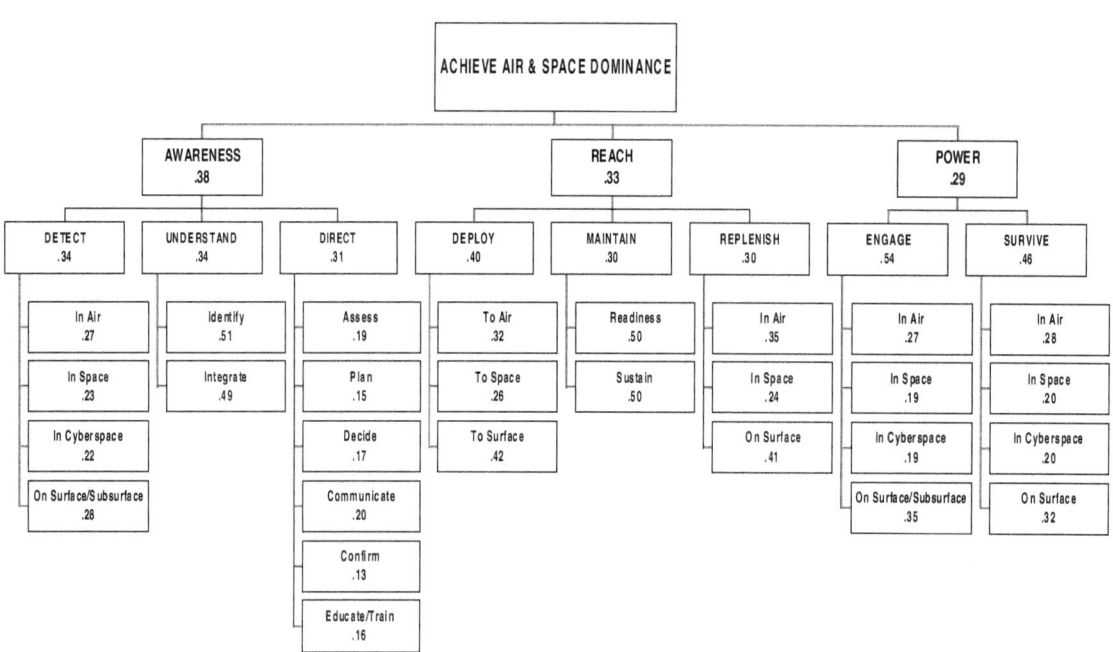

Figure C-5. AU Team Weights - Khan Future

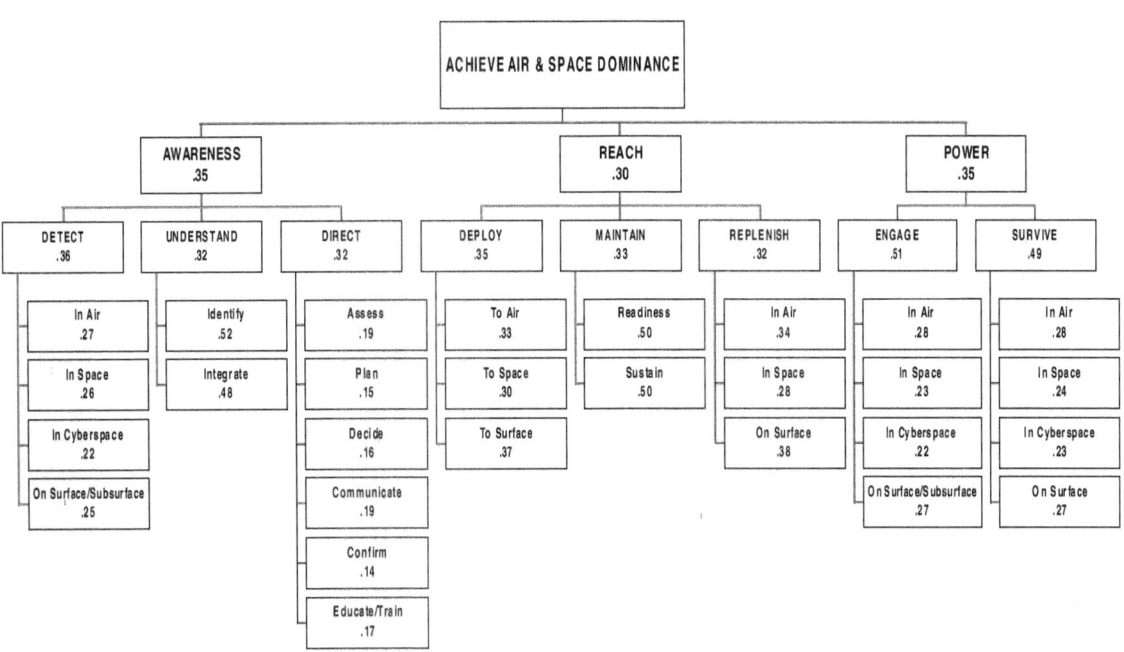

Figure C-6. AU Team Weights - 2015 Future

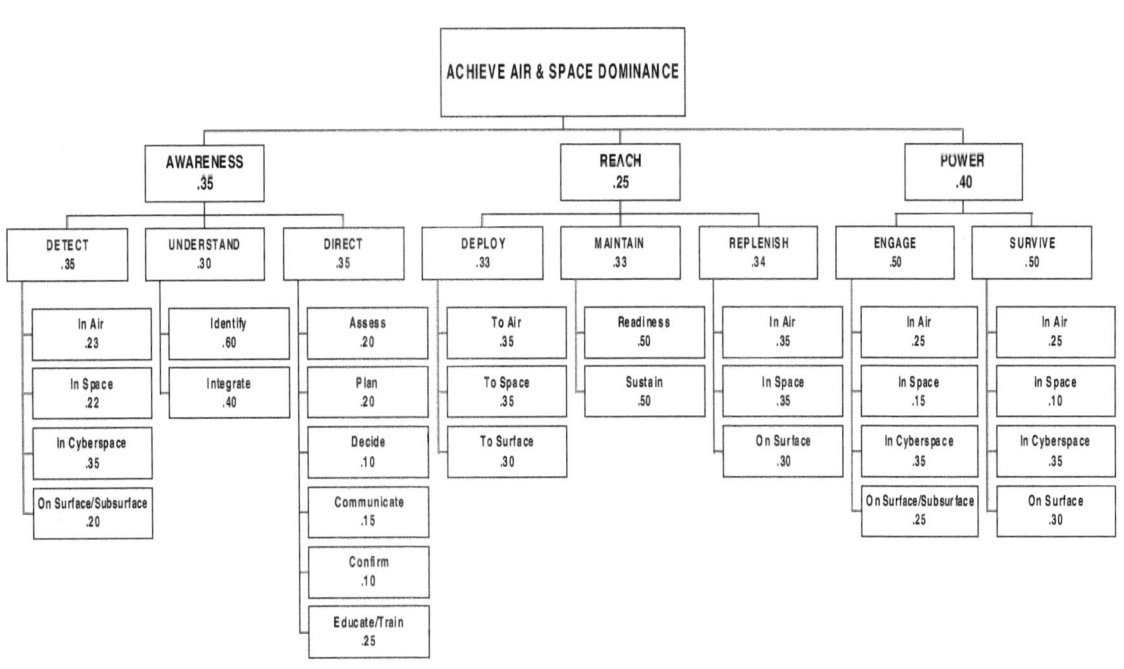

Figure C-7. Alt Futures Weights - Halfs Future

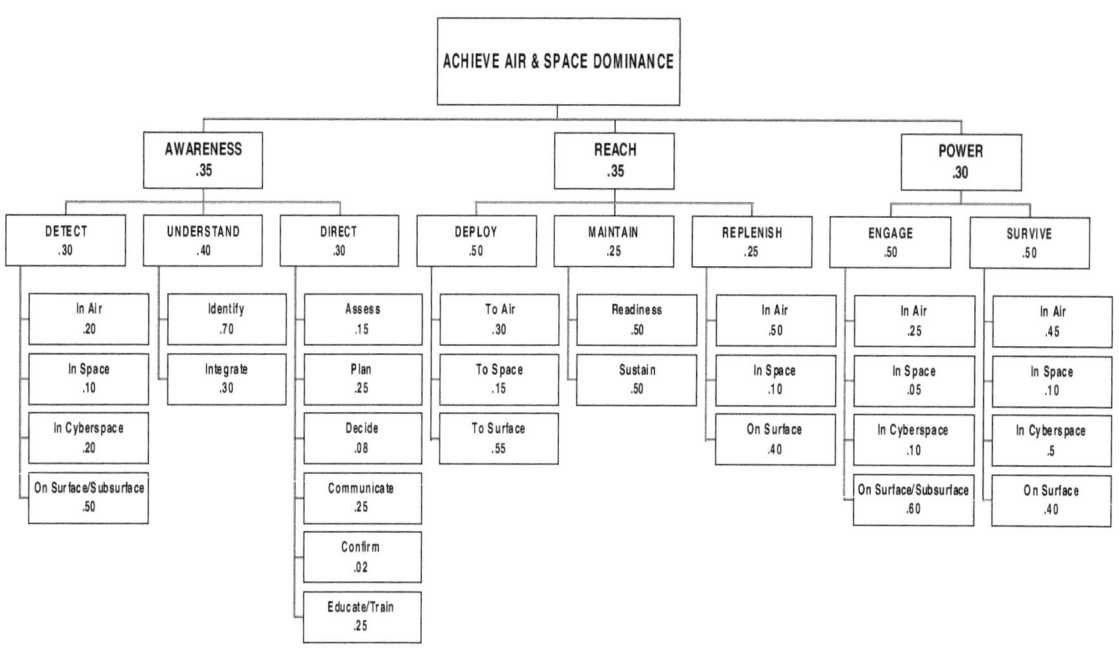

Figure C-8. Alt Futures Weights - Gulliver's Future

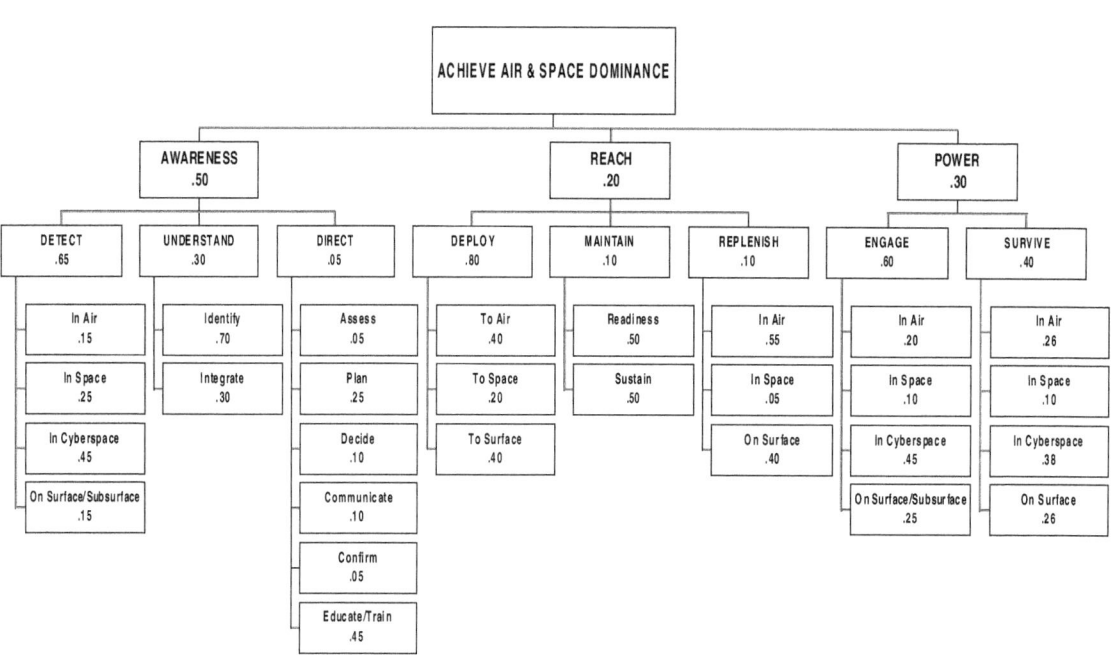

Figure C-9. Alt Futures Weights - Zaibatsu Future

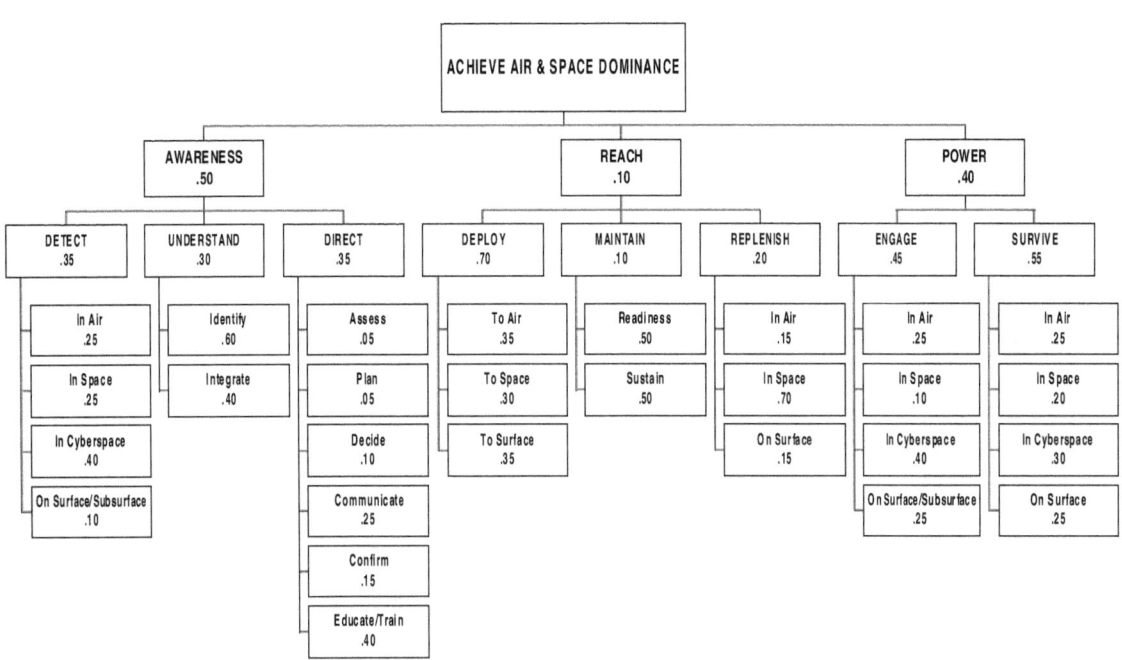

Figure C-10. Alt Futures Weights - Digital Future

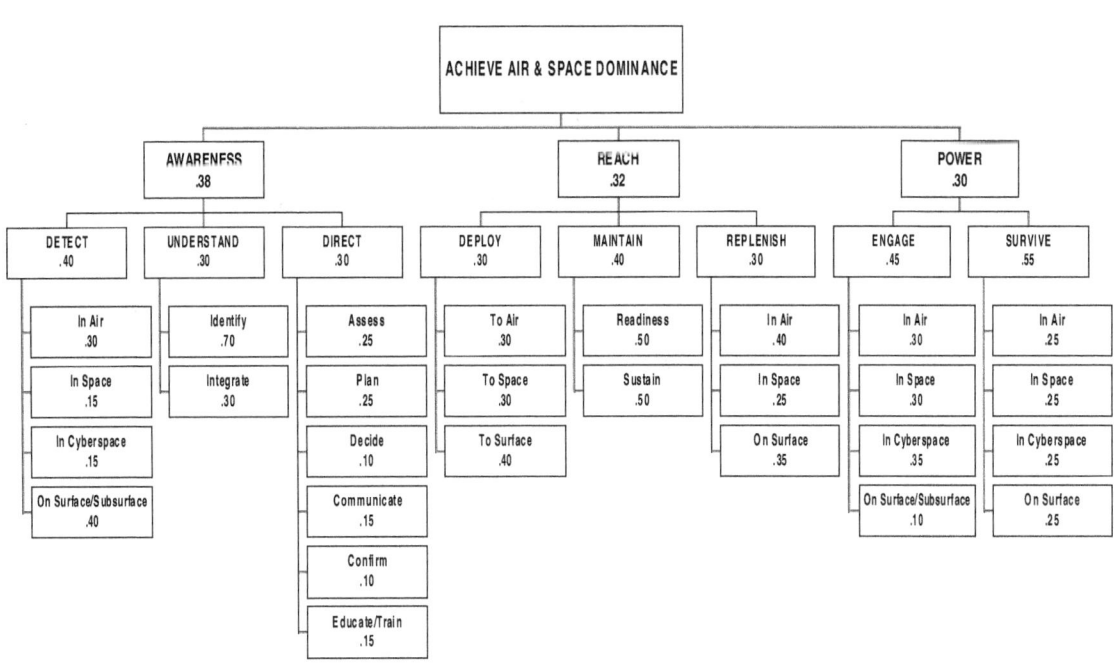

Figure C-11. Alt Futures Weights - Khan Future

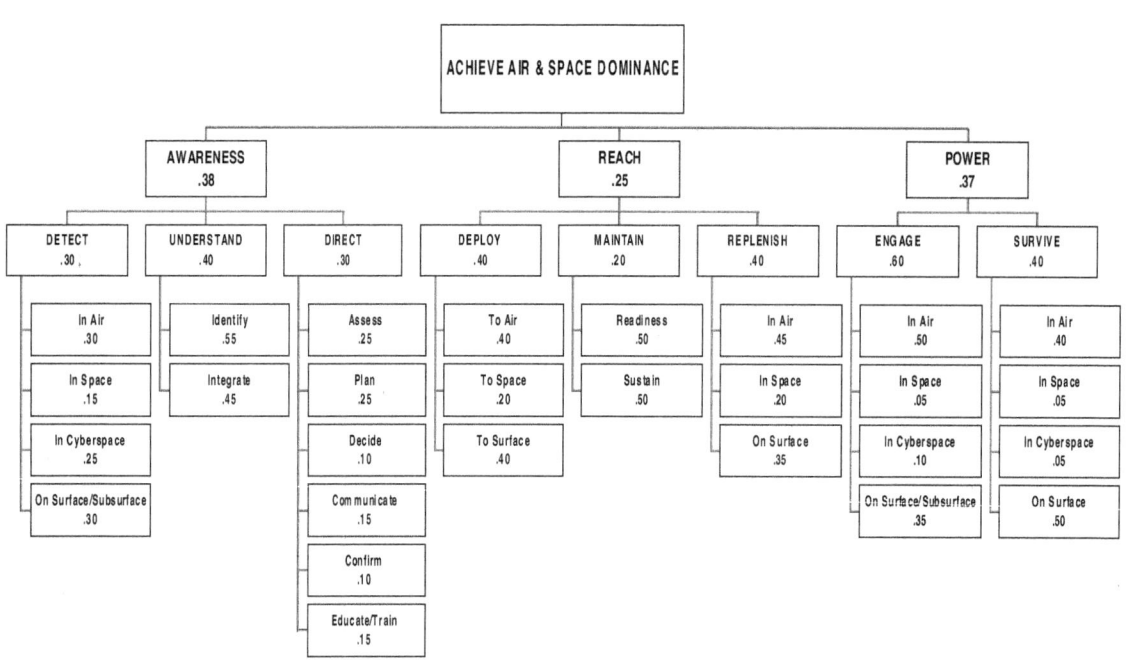

Figure C-12. Alt Futures Weights - 2015 Future

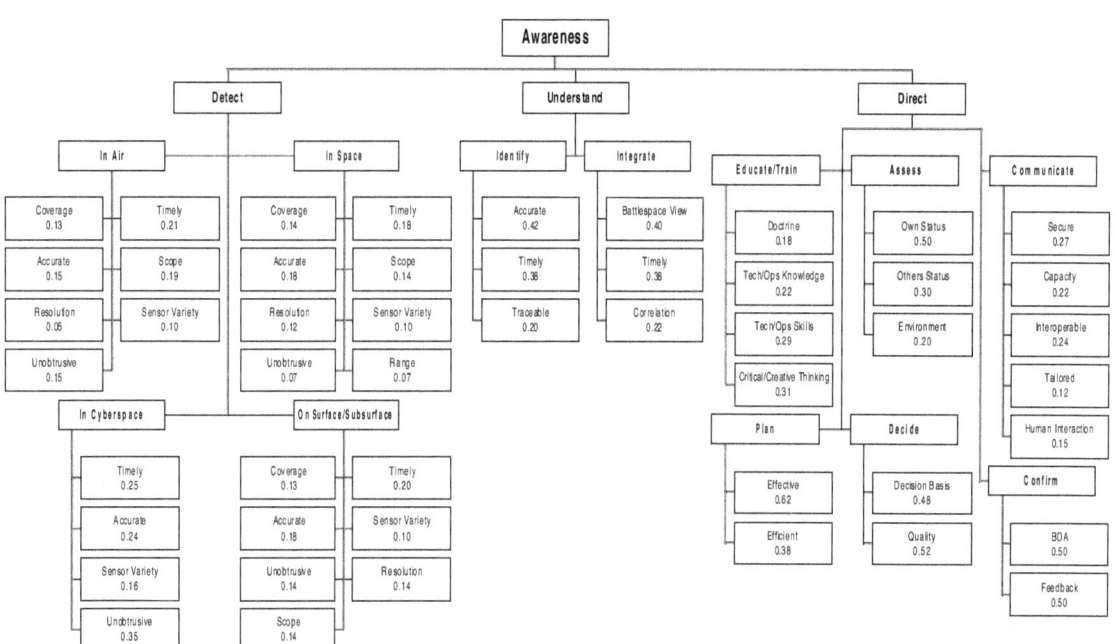

Figure C-13. *Awareness* Force Quality Weights

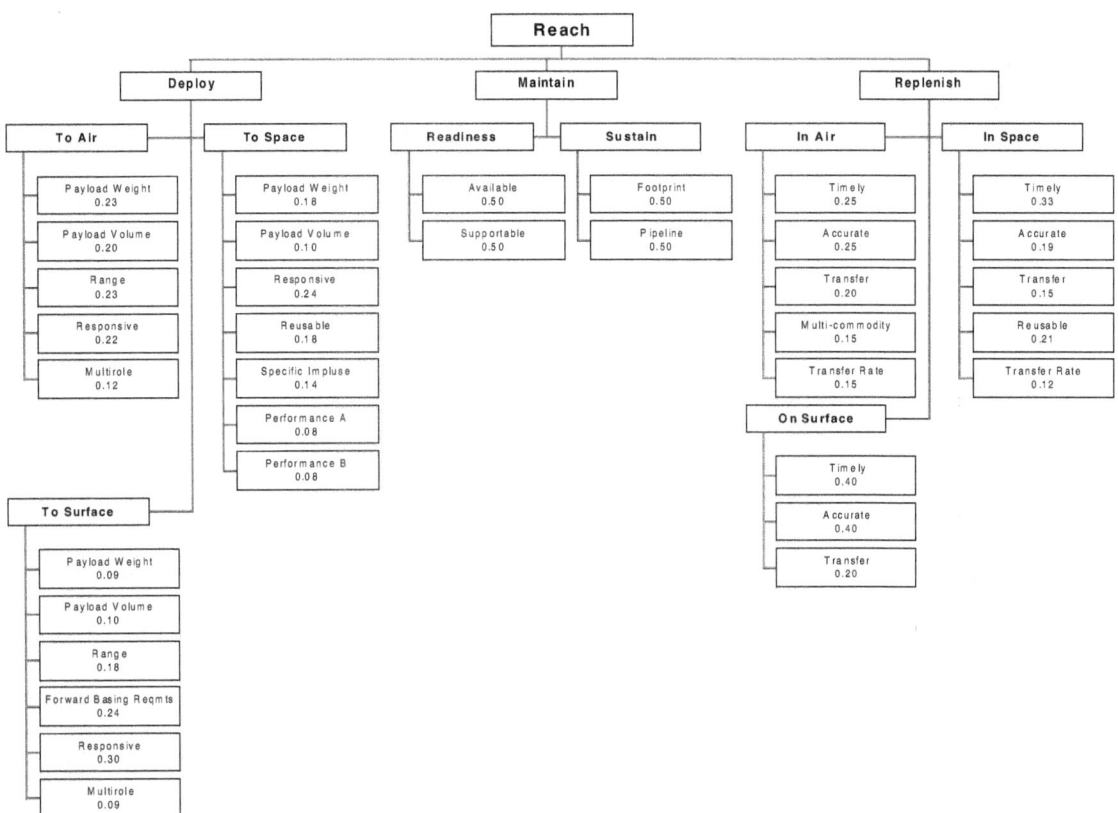

Figure C-14. *Reach* Force Quality Weights

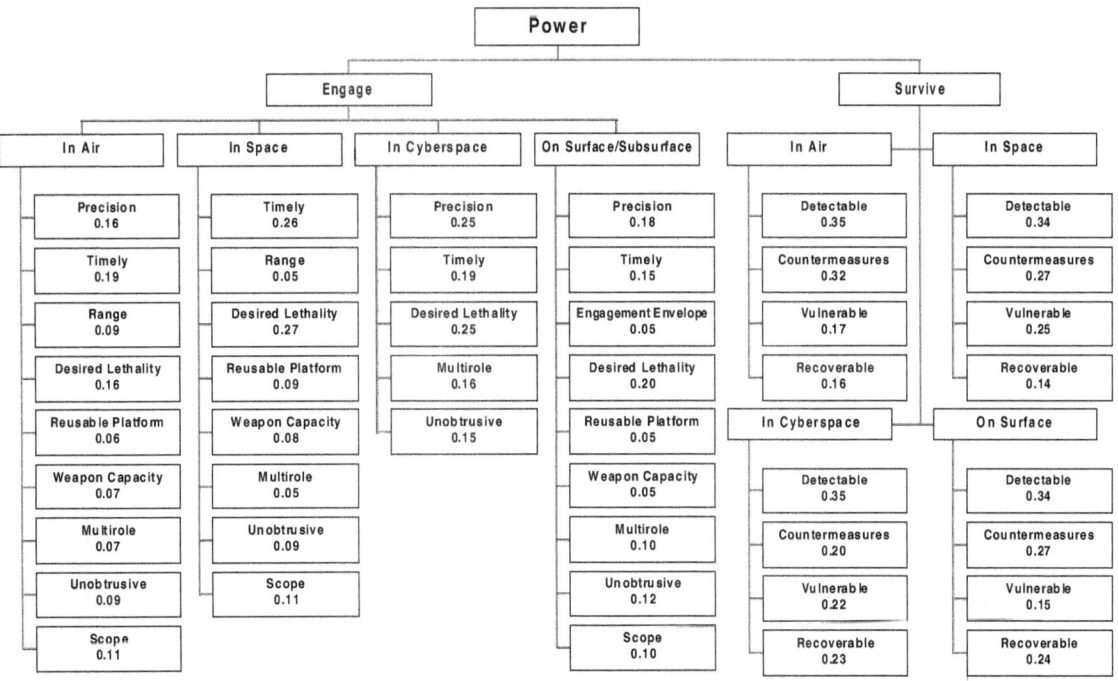

Figure C-15. *Power* Force Quality Weights

Appendix D

Technology Model

This appendix contains a description of the leveraging technologies identified during the course of system analysis. The technology names, numbering convention, and descriptions contained in the Military Critical Technologies List (MCTL) served as the basis for the 2025 Technology Model. Descriptions of three technology areas not fully developed in the MCTL (artificial intelligence, other propulsion, and other sensors) were developed by the Analysis team and inserted into the 2025 Technology Model (fig D-1 and fig. D-2).

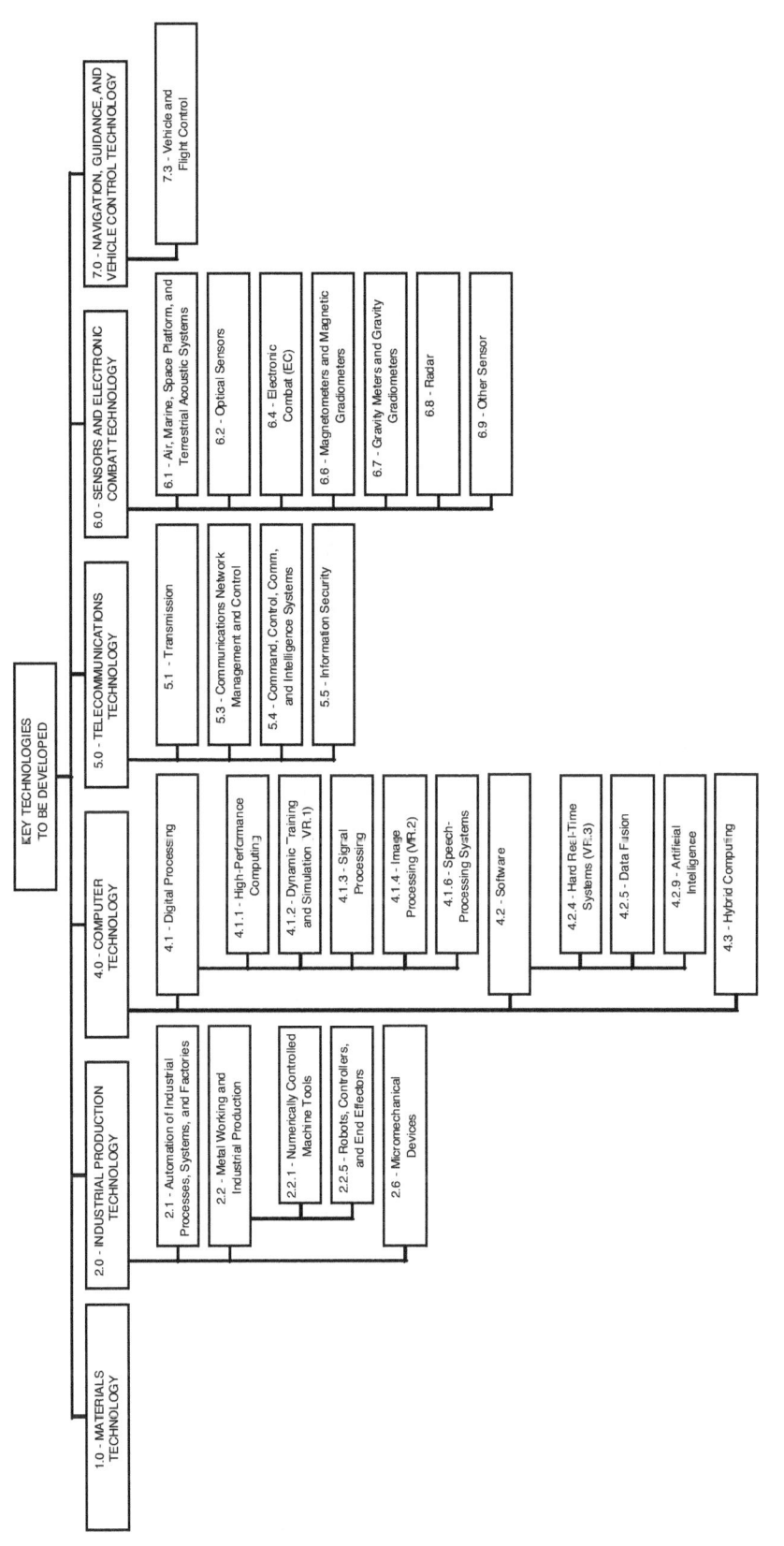

Figure D-1. Technology Model - Part I

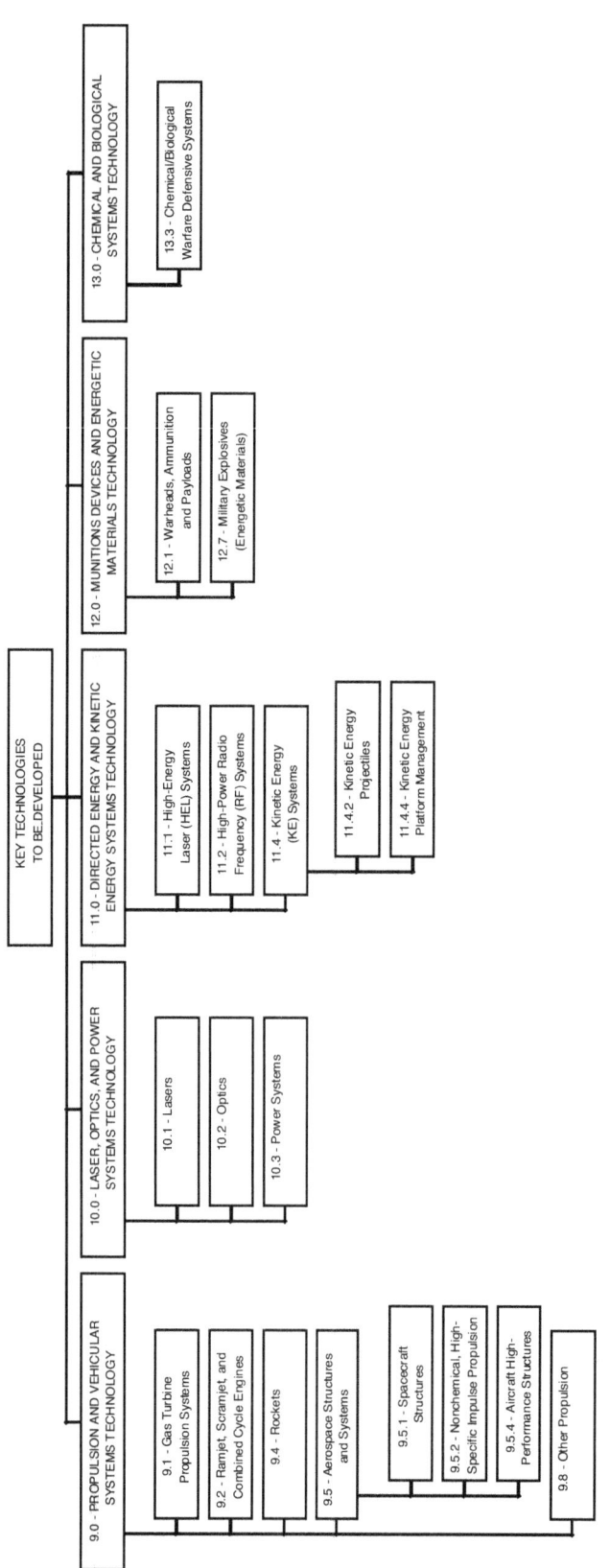

Figure D-2. Technology Model - Part II

Level 1 Technologies

1.0 Materials Technology

This section covers specific dual-use, multiapplication materials and technologies. The metallic, ceramic, and composite materials are principally related to structural, load-bearing functions in aircraft, spacecraft, missiles, surface, and undersea vehicles and in propulsion devices. The selected polymeric materials provide seals and sealants for containment of identified fluids and lubricants for various vehicle and devices, as well as the critical matrix for composites. Selected specialty materials provide critical capabilities that exploit electromagnetic absorption or magnetic or superconductivity characteristics.

2.0 Industrial Production Technology

Production technology includes the facilities, equipment, tools, information systems, and infrastructure for production enterprise. Elements of this technology include automation of part production and assembly, equipment and process tools, real-time planning and control systems, modeling and decision technologies, and the underlying information systems that enable use of the technology. The equipment and technologies of concern in this section support the conversion of raw and semifinished materials into products or components thereof through a sequence of steps which include production design and engineering, manufacturing, and quality inspection and testing. The steps involve computer aided design (CAD)/computer aided manufacturing (CAM), complete factory design, and automation as well as needed technologies for the development and production and integration of the implementing machines, equipment, tools, and software. Technology, software, equipment, and plants for material removal, forming, joining, inspection, coating, measuring, and other generic development and production processes and important components for these production equipments are included. Bearings which are key components not only for production equipment but also for hardware are covered here because of their generic nature and unique industrial base.

4.0 Computer Technology

Existing computer development and production technologies span a wide segment of products, some of which are not practical to control. In this category are products known to be obtainable from multiple sources that are readily accessible outside the US. These products may be decontrolled, but the development and production technologies of embargoed computer products should remain under control. Computers in this section are broadly categorized as digital, hybrid, and advanced. The digital computer classification encompasses primarily those computers whose performance capability exceed certain minimums, including those optimized for signal processing and image processing. A discriminating criterion for this entire category is the ability to do sensor-based processing in real time. Hybrid computers are the amalgam of analogue and digital computers that collectively perform certain computational tasks much faster than on digital computers. Advanced computers are based on technology and structure differing from both digital and hybrid computers. There performance exceeds that of either digital or hybrid computer technology.

5.0 Telecommunications Technology

This section covers technology for telecommunication equipment used to transfer voice, data, and record information. It encompasses technologies for transmission equipment operating over wire, coaxial, or optical fiber cable or using electromagnetic radiation; radio-transmitting and receiving equipment, including radio telephone, cellular radio, radio relay and satellites; optical fiber and accessories for information transmission and optical fiber manufacturing equipment; and stored program controlled circuit and packet-switching equipment and networks. Information being exchanged is predominantly in digital form for text, graphic, video, and databases. This format permits the application of security as required. The cryptographic equipment to ensure secrecy for communications, video, data, or stored information and the software controlling or computers performing the functions of such cryptographic equipment are covered.

6.0 SENSORS AND ELECTRONIC COMBAT TECHNOLOGY

This section covers all sensor types (except chemical and nuclear) that are of military interest. The types include technologies for acoustics, optical sensors, cameras, radar and identification, gravity meters, magnetometers and associated gradiometers. This section contains a subsection on electronic combat technology since these systems are so closely associated with sensors.

7.0 Navigation, Guidance, and Vehicle Control

This section encompasses technologies for both autonomous and cooperative positioning, coordination, and control of military force elements. Included are technologies for flight management, vehicle guidance, and control. In essence, these technologies are closely coupled and overlap depending on application. Navigation is defined as obtaining the present condition or state of the vehicle from sensed values of position and motion. Navigation is subdivided into inertial-based and position (radio)-based subsections. Guidance systems integrate these conditions and produce vehicle control responses.

9.0 Propulsion and Vehicular Systems Technology

Specific technologies and hardware discussed relate to aerogas turbines and diesel engines, particularly those that improve fuel efficiency and power output per pound of engine weight or engine cube and extend engine life; the leading edge technologies associated with such near-earth, exoatmospheric super, and hypersonic flight propulsion systems as RAMJET, SCRAMJET, and combined cycle engines and liquid and solid rockets; and, the critical technologies associated with the vehicle structures and systems for which the above propulsion systems are usually provided.

10.0 Laser, Optics and Power Systems Technology

This section covers dual-use technology for lasers, optics and power systems. It covers both high-energy laser (HEL) and low-energy laser (LEL) systems. Optical technologies encompass optical materials, optical filters, optical components for both LEL and HEL applications, optical computing and photonics for instrumentation and signal processing, and large optics for space surveillance applications. Nonlinear optics for beam phase conjugation and image enhancement applications, shared apertures for both directed energy/high energy lasers (DE/HEL) weapons and space tracking applications, as well as cooled laser optics for both active and passive cooling applications also are covered. Power systems covered include energy conversion and storage, generation, conditioning, control, and distribution of electrical power for both one-shot, continuous, and pulsed applications.

11.0 Directed Energy and Kinetic Energy Systems Technology

Directed energy (DE) and kinetic energy (KE) technologies can incorporate one or more of the following: laser systems; charged and neutral particle-beam systems; radio frequency (RF) systems, or kinetic energy systems. This section includes technologies required to generate electromagnetic radiation or particle beams and to project that energy or particle to a target where it will perform interference with system functions, destruction, degradation, or system abort. This section also covers technologies required to impart a high velocity to a mass such as those used in KE systems and direct it to a target.

12.0 Munitions Devices and Energetic Materials Technology

Munitions devices covered by this section are limited to those for conventional systems, to the integration of munitions/weapons subsystems into effective operational systems, and to components that are acted upon by energetic materials or that act on energetic materials. Also covered are energetic materials and fuels technology and the precursors that are necessary for their manufacture. *Energetic materials* is the collective term for high explosives, propellants, and pyrotechnics.

13.0 Chemical and Biological Systems Technology

Chemical and biological technologies cover the following areas: (1) chemical and biological systems and technologies for developing toxic chemical, biological, and toxin agents and their dissemination and delivery; (2) items for detecting, warning, and countermeasuring and (3) a specific set of advanced biotechnologies that are of explicit military importance and embody critical or unique capabilities.

Level 2 Technologies

2.1 Automation of Industrial Processes, Systems and Factories

Modern industrial facilities are applying automation technologies increasingly to achieve integrated factory control systems. The technologies of concern in this section support a distributed or hierarchical approach to factory automation consisting of the development, use, and production of key-technology elements and the integration and implementation of these elements to provide automated control of industrial systems. A primary vehicle for transfer of critical technology (know-how) addressed in these sections is access to extractable information in software. Such know-how includes engineering design and manufacturing process information embedded in software (e.g., control algorithms, models, and knowledge-based CAD/CAM products) embodying critical technologies. This information can take a variety of forms, including empirically validated engineering design databases or expert systems designed for CAD/CAM or industrial automation applications.

2.2 Metal Working and Industrial Production

This subsection covers the technology for the general category of machinery and equipment used for the production of systems and components. The equipment described are needed to produce other production equipment, as well. The individual pieces of machine tools, machinery, and process equipment provide the foundation for the manufacturing level of a plant. Most incorporate or are equipped with electronic controllers to assist in providing accuracy, flexibility, and speed to the production process. The individual machines may be integrated into work stations or cells to communicate and be integrated into higher level industrial automation systems.

2.6 Micromechanical Devices

This subsection covers the production and technologies for the manufacture of micromechanical devices (also known as micromachines, microrobots, and microsensors) and their components. Focus is on scale and miniaturization of micromechanical technology for structural components using technologies characteristic of microelectronic production. The three major areas of concern include (1) manufacture and inspection of parts with nanometer range tolerances; (2) manufacture and inspection of integrated micromechanical and microelectronic systems with nano meter range tolerances; and (3) assemblies of these parts into subcomponents and systems which require submicrometer and finer tolerances. Current state-of-the-art machine tools can make/produce parts with tolerances (accuracies of 0.5 to 1 micron (1,000 nanometers). Specialized precision machine tools can produce parts in the 200-nanometer range. Special polishing processes can produce surface finishes of 10 nanometer surface roughness or better. Technology and equipment for producing such large parts as laser mirrors, other optics, molds for precision casting, suspension systems, hydraulic system components, accelerometer, gyro, missile guidance, and inertial navigation components which require nanometer tolerances for surface finish and form fall within the scope of this section.

4.1 Digital Processing

High-performance digital computing is an area of critical concern. The development, production, service, and operational use of high-performance digital computing are also of concern. In addition to addressing these functions, the section also focuses on specific designs, key storage equipment and, critical applications.

4.2   Software

The broad categories in this subsection address software directly applicable to military activities; software engineering used to support the development, operation, and maintenance of other software; and general-purpose software that has a dual role in supporting military data processing.

4.3   Hybrid Computing

Hybrid computing technology provides the functional integration of digital and analogue processing primarily for dynamic simulation of complex physical systems. Dynamic simulation (hardware in the loop) is an essential step in the development of virtually all tactical and strategic guided weapons. Dynamic scene generators that provide computer generated imagery for the testing and training of infrared seeking sensor systems are integral to the development of guided weapons employing such features. Hybrid technology realizes both the inherent speed of analogue processing and the programmability and accuracy of digital processing required for weapon design and testing.

5.1   Transmission

This subsection covers technology for telecommunications transmission equipment and components used for transfer of voice, data, record, and other information by electromagnetic means either through atmospheric, exoatmospheric, or subsurface (water) media or through metallic or fiber optic cable. Controllability is of concern since telecommunications technologies below the control level may suffice for transfer of the telemetering information.

5.3   Communications Network Management and Control

Communications technology covered in this item is highly dependent on the automation of the monitoring and controlling functions within the communications network. The monitoring and controlling functions are combined in separate systems capable of working over a widely dispersed geographical area with equipment using various transmission media and switches using common channel signalling. These systems provide a centralized control capability to configure transmission equipment to optimize networks for loading and failures; configure switches and routers to optimize the call distribution within a network; and alter network configurations and routing information to provide special network service features. The information transferred using transmission control protocol/internet protocol (TCP/IP) or similar protocols ranges from basic integrated services digital network (ISDN) and fractional multiplexers up to 45 Mbit/s and network access controllers operating at a digital transfer rate up to 33 Mbit/s with local area networks (LAN)/wide area networks (WAN).

5.4   Command, Control, Communications, and Intelligence Systems

Integrated C^3I systems are fabricated combinations of platforms, sensors and weapons, software and data-processing equipment; related communications subsystems; and user-system interfaces specifically designed for the control of US armed forces and weapons systems. Command, control, communications, and intelligence systems are integrated combinations of military command information processing, communications network, and intelligence-gathering subsystems (including surveillance, warning, and identification subsystems) that make up the US C^3I systems. These combined technologies support US authorities at all echelons with the integrated C^3I systems that provide the timely and adequate data required to plan, direct, and control US military forces and operations in the accomplishment of their missions.

5.5   Information Scurity

Information security includes the means and functions controlling the accessibility or ensuring the confidentiality or integrity of information and communications, as well as the availability of resources, excluding the means and functions intended to safeguard against malfunctions.

6.1 Ar, Mrine, Sace Patform and Terrestrial Acoustic Systems

Most marine-sensing systems use sonars which employ acoustic signals (sound waves) to locate underwater objects and features. Sonars are termed active when sound is generated by the system for the purpose of echo ranging on a target and passive when listening to the sound radiated by the target. Sonar systems are the principal sensors used for anti-submarine warfare (ASW), mine warfare, and deep-sea salvage for the purpose of detecting, classifying, identifying, locating, and tracking potential underwater targets. Additional uses are undersea mine and torpedo homing and activation, depth sounding and bottom mapping. Sonars are used commercially for locating fish and other objects in the ocean, for seismic exploration at sea, for petroleum and mineral exploitation, and for academic studies. Most terrestrial (land-based) seismic systems employ sound waves to locate features in the earth's crust for geophysical prospecting. All acoustic seismic systems are active. Acoustic vibrations are also a critical issue for space platform stability. Aircraft and space sensors and optical systems require stabilization from acoustic coupling due to propulsion and power generation equipment acoustic noise.

6.2 Optical Sensors

This subsection covers equipment and components that are sensitive to emissions in the electro-optic (EO) portion of the electromagnetic spectrum, (ultra-violet, visible, and infrared in micrometers: UV 0.01-0.4, VIS 0.4-0.8; IR 0.8-30.0) and support equipment therefor. Applications include seeker heads, night vision imaging, intrusion detection, medical diagnosis, and earth resource analysis.

6.4 Electronic Combat

This subsection includes electronic support measures (ESM), electronic countermeasures (ECM) and electronic counter-countermeasures (ECCM), all of which are part of electronic combat. ESM involve actions taken to search for, intercept, identify, and locate radiated electromagnetic energy. ECM involves actions taken to prevent or reduce an enemy's effective use of the electromagnetic spectrum. And ECCM involves actions taken to ensure friendly effective use of the electromagnetic spectrum despite the enemy's use of ECM.

6.6 Magnetometers and Magnetic Gradiometers

Magnetic sensor systems detect and display the presence of a magnetic field and measure its magnitude and/or direction. Some magnetic sensors are sensitive to their orientation with respect to the vector components of the magnetic field in which they find themselves; others are not. Some are capable of measuring the absolute level of an ambient magnetic field; others can only measure variations. Magnetic sensor systems are often configured to detect the spatial variation of the magnetic field intensity, that is, the gradient of the magnetic field intensity. In this mode, they are called magnetic gradiometers. Magnetic gradiometers can consist of two magnetometers or consist of a single intrinsic sensor. Compensation for background noise and processing of the sensor data are required to obtain optimal performance in a given application.

6.7 Gravity Meters and Gravity Gradiometers

Gravity meters can be used in a static or moving base mode to measure gravity magnitude. The gravity gradiometer, used in the same modes as above, is used to measure gravity gradients. Gravity data is used to estimate vertical deflection and other gravity field components. An absolute gravimeter is usually operated at a fixed installation by the accurate (better than 10^{-8} seconds) timing of a falling weight or of a swinging pendulum. Due to the improvement of timing devices and instrumentation, the freely falling body method is currently favored. Distance accuracy measurement of better (or less) than 0.5 micrometer is needed to obtain an accuracy of 1 mGal with a fall of 1 or 2 m. Previously, a large number of oscillations of a pendulum was the preferred absolute gravity-measuring method. Astrolabes are used to measure astronomic positions which in conjunction with geodetic position are used to compute deflection of the vertical. Whereas gravity mapping is required over a significant area to calculate deflection of the vertical, the astrolabe allows determination at a single point. These sensors can be used commercially to assist in the exploration for such natural resources as oil, gas, or minerals. Since gravity and spatial accelerations are not separable, inertial navigation and guidance systems require compensation based on knowledge of the gravity field. Gravity meter accuracy is essential for siting and initializing ballistic missiles and other long-range unaided inertial military system applications.

6.8 Radar

This subsection covers RF and laser radars, including identification, recognition and classification technology. Radar has many military and civil/commercial uses. Applications of radar include surface and airborne surveillance, fire control, air traffic control, and meteorological analysis. Identification and classification technologies involving noncooperative targets are a matter of heavy development effort with some real promise of success as data-processing speeds and digital memory hardware ramifications allow massive processing in small packages.

6.9 Other Sensor

This subsection covers sensor types not specifically covered in other subsections.

7.3 Vehicle and Flight Control

This item covers the technologies that collect, integrate, and analyze data from multiple on-board aircraft sensors and instruments, through use of high-speed buses and data-processing units and provide military aircraft, space vehicles, and ground combat vehicles with flight movement control systems that optimize their performance for specific mission profiles. These guidance and control management systems convey the directions of the operator by providing the optimum control information to the vehicle for the situation at the time of operator input (avoiding over or under control). Since vehicle and flight control systems can only operate with the real-time collection, analysis, and feedback of data, the systems depend upon technologies associated with state-of-the-art sensors.

9.1 Gas Turbine Propulsion Systems

This subsection covers gas turbine development, production. and use technologies, particularly those related to extending engine life and performance. This section also includes advanced engine test facility technologies and hardware and full authority digital engine controls, diagnostic equipment, engine structures, and materials. Technologies of concern support higher thrust-to-weight ratios and higher internal operating temperatures and variable cycles, with increased fuel efficiency and longer operating life.

9.2 RAMjet, SCRAMjet, Combined Cycle Engines

Ramjet/Scramjet/combined cycle propulsion systems provide flight speed capability for air-breathing propulsion systems to Mach 25. Combustion can take place either subsonically (Ramjet), supersonically (Scramjet), or both (i.e., dual-mode subsonic/supersonic combustion). Two types of Ramjet engine configurations are the podded Ramjet and the integral rocket Ramjet (IRR). Many types of combined cycle engines are being studied, however, the most common types include the air-turborocket or air-turbojet, turboRamjet, liquid air cycle engine, dual-mode Ramjet/Scramjet engine, and turborocket Ramjet engines. Several operational Ramjets currently have been limited to military applications and flight speeds up to Mach 4. There are no currently operational combined cycle engine types in the United States. Future applications of the Ramjet and combined cycle engines include hypersonic missiles capable of flight speeds up to Mach 6 and hypersonic aircraft and advanced space launch vehicles for either military or commercial use capable of flight speeds up to Mach 25. Technologies of concern are those for the design and production of the Ramjet and technologies unique to the integral rocket Ramjet and combined cycle engines. This item covers both fixed- and variable-area inlets for Ramjet and combined cycle engines, including installed inlet performance analysis of relative mass flow/pressure recovery and internal flow field analysis; high-density fuels and fuel delivery systems used to permit Ramjet and Scramjet vehicles to operate up to Mach 6-10. It also includes the tank/vessel structure, insulation, valve/control devices for fuel metering, and the fuel injection devices. Finally, this item includes flame-holder/combustion aids, insulation systems, igniter systems, transition devices (from rocket to Ramjet operation), and nozzle design. Combustor insulation includes both short-life ablator systems and longer life nonablating systems, as well as systems which include fuel or air cooling.

9.4 Rockets

The technologies discussed here focus on the propulsion of rockets and missiles by either solid, liquid, or hybrid propulsion motors. The technologies of concern are those associated with providing more efficient propulsion through better propulsion control, more lightweight motor hardware, more efficient subsystems, and better development and production processes.

9.5 Aerospace Structures and Systems

Advances in structures technology are expected to improve aircraft and spacecraft performance significantly. Incorporation of new materials and design concepts in aircraft structure will expand aircraft operating envelopes, reduce observability, and improve survivability to ballistic and laser threats. Incorporation of new materials and structures concepts into spacecraft will significantly reduce weight, provide increased precision for greater mission performance, reduce cost, improve producibility, and contribute to improved survivability to natural and man-made threats. New materials offer high strength to weight, stiffness to weight, and greater resistance to fatigue and corrosion; high-temperature durability; and increased toughness, moisture resistance, and reformability. Structures-related technologies are vital to systems which play major roles in strategic and tactical warfare missions. Because many problems and technologies related to survivability apply to ground vehicles as well as to aerospace vehicles, survivability of both sets of vehicles is covered in this section.

9.8 Other Propulsion

This subsection covers propulsion types not specifically covered in other subsections

10.1 Lasers

Technologies applicable to the development and production of lasers and laser systems in the infrared (IR), visible, and ultraviolet (UV) regions of the electromagnetic (EM) spectrum (0.3-0.01 to 30 μm) capable of achieving significant levels of energy or power are covered. Lasers consist of the laser hardware (the device) and laser medium (host material), mirrors, and other optical components that form the laser oscillator cavity. Lasers may operate in a continuous, repetitive, repetitive burst, or single-pulsed mode depending on the application and requirements. Laser systems incorporate such components as amplifier stages, frequency conversion components, Raman cells, multiple wave mixing components, or other major elements in addition to the laser oscillator.

10.2 Optics

Optical technologies encompass optical materials, optical filters, optical components for both low-energy laser (LEL) and high-energy laser (HEL) applications, optical computing and photonics for instrumentation and signal processing, and large optics for space surveillance applications. Nonlinear optics for beam phase conjugation and image enhancement applications, shared apertures for both directed energy/high-energy lasers (DE/HEL) weapons and space-tracking applications, as well as cooled laser optics for both active and passive cooling applications also are covered.

10.3 Power Systems

This subsection covers technologies for nonnuclear energy conversion and electrical power generation, energy storage, and power conditioning and pulsed power systems and power control and distribution. Technologies related to devices and components and thermal management and system integration technologies are also included. Well-regulated and high-power-density power supplies are a key supporting technology for reliable computing and in telecommunications and information systems.

11.1 High-Energy Laser Systems

This subsection includes those technologies required to generate laser beams and project them to a target where they will perform destruction, degradation, or mission abort. Mission abort includes both temporary interference with the ability of the system to function as well as permanent degradation which prevents proper system function. This class of technology is called high-energy laser-directed energy. Included are those technologies required to generate, project, and couple the beam while tracking the target and aim point and propagating through the atmosphere or a space environment. This section also includes the technologies required to implement and integrate them into a DE system. HEL systems are capable of generating intense beams (20 kW or greater average power, 1 kJ or more energy per pulse) in the infrared, visible, and ultraviolet regions of the electromagnetic spectrum (0.3 to 30 μm). Depending upon the laser type, the system may operate in a continuous, repetitive burst, or single-pulsed mode. HEL systems consist of the laser device, mirrors and other optical components that form the laser cavity. HEL/DE systems consist of the laser plus a control subsystem for directing the beam to a target and other optical components used to stabilize or improve the laser beam quality.

11.2 High-Power Radio Frequency Systems

High-power RF systems project intense power in either pulsed or continuous waves. In some contexts, such technology and systems are known as high-power microwave (HPM). These systems consist of sources capable of generating sufficient RF power, components for modulating the power, and antenna arrays which can direct the energy to a target. Current frequencies of interest lie between 0.1 and 1,000 GHz. Power levels of interest include peak power of 100 megawatts or more, single-pulse energy of 100 J or more, and average power of more than 10 kW. Sources which can operate in the aforementioned regimes and are amenable to weaponization are critical to the development of RF weapons systems. The principal technologies related to HPM are pulsed power and those concerned with the protection of targets from RF radiation.

11.4 Kinetic Energy Systems

Kineticenergy technologies of interest are those required to propel projectiles to higher velocities (greater than 1.6 km/sec) than are typical of projectiles from conventional gun or rocket systems and to obtain an appropriate combination of properties including shape, size, density, and ductility at that velocity. Technologies for precision pointing, tracking, and launch and for management of launch platforms are also of concern.

12.1 Warheads, Ammunition and Payloads

This subsection covers the technology, materials, and equipment necessary to develop, produce, and integrate conventional, improved conventional, precision-guided, or smart munitions for air-, sea-, or ground-launched systems projectiles, mines (sea and land), bombs, fuel-air munitions, mortar rounds, torpedoes, and missile and rocket warheads.

12.7 Military Explosives (Energetic Materials)

This subsection covers technology to develop and produce the ingredients of high-explosive, propellant, and pyrotechnic formulations and to use the ingredients in the development and production of militarily useful formulations. Materials, equipment, and processing principles necessary for manufacturing facilities draw on developments within the commercial chemical processing industry, and some are widely available chemical products. However, the listed materials are primarily, if not uniquely, used for the manufacture of warheads, propulsion systems, and the guidance and control systems of munitions controlled under the US munitions list (USML), international munitions list (IML) and the missile technology control regime (MTCR). After the materials (ingredients) have been produced in bulk, technology is required to develop, blend, and formulate compositions to give them the characteristics necessary for particular applications; to fabricate charges or to load them in selected munitions systems; and to test the charges or systems to assure that the performance objectives have been met. Coverage is limited to the listed substances and to other substances or compositions that meet specified performance criteria. It includes both theoretical (computational) and empirical methods of defining required compositions and properties and chemical and physical processes for incorporating these in military systems. Commercial application of the formulation, fabrication, and test technology is limited, but the principles and the equipment, facilities, and software also can be used to optimize and produce commercial explosive products. The development of formulations and the production of charges are also necessary for commercial devices that are used in blasting, oil exploration, and oil well stimulation; for explosive-forming of metals and compacts; and for space vehicles.

13.3 Chemical/Biological Warfare Defensive Systems

This subsection covers the technologies and materials relative to defend against chemical and biological warfare (CBW) and biopolymer-based detection and warning systems. Defense against CBW entails detection and warning, individual and collective protection systems, and countermeasures, including decontamination.

Level 3 Technologies

2.2.1 Numerically Controlled Machine Tools

This item includes numerically controlled (NC) units, motion control boards, NC machine tools for turning, milling, and grinding, nontraditional machine tools, and components for material removal machines. NC machining provides the ability to produce repeatedly parts of accurate dimensions containing complex geometry, allowing designers to specify components closer to desired functional shape with fewer manufacturing concerns. Repeatability and accuracy of parts lead to the higher performance ratings and endurance of US equipment.

2.2.5 Robots, Controllers and End Effectors

This item covers industrial and military smart robots, controllers and end effectors capable of employing feedback information in real-time processing from one or more sensors to generate or modify programs or program data. These robots are multifunctional manipulation devices capable of positioning or orienting parts, tools, or other devices through variable movements in three-dimensional space and thus would have three or more closed or open-loop servo devices. They have user accessible programmability by means of teach/playback or off-line computer or programmable logic controllers. This technology is critical to the delivery of essential numbers of required quality systems to the battlefield. It applies in industrial processes and in the battlefield. Robotic technology is used in industry to provide the quality and quantity of goods needed while providing for personnel safety and industrial flexibility to meet changing needs. Radiation-hardened robots are used in nuclear reprocessing and nuclear production reactor activities. They may be used in nuclear facilities to reduce occupational radiation exposures.

4.1.1 High-Performance Computing

The highest performance digital computers currently in use are composed of vector and scalar processors maximized for performance with some achieving clock cycles under five nanoseconds. Logic and memory device technology appears to limit individual processor clock cycle advances to approximately one nanosecond. Interconnecting a multiplicity of these units into coupled parallel combinations continues to yield performance gains with 16 vector processors being generally the maximum size envisioned. A significantly different parallel processing architecture involves the interconnection of conventional digital processing elements. Medium parallel combinations range from an interconnection of 16 to 100 processors using processors composed of work stations or low-end technology devices. Massively parallel computers range from 100 to 100,000 and beyond processor combinations that yield high theoretical performance but encounter difficult interconnection, programming, and use problems. Visualization is a presentation technique aiding in the extraction, analysis, and interpretation of voluminous data sets required for solutions of problems in such fields as astrophysics, geophysics, fluid dynamics, chemistry, structural analysis, and high-energy physics. For large data sets, visualization can only be achieved by a computational environment containing a super computer for modeling, simulation and analysis, a large main frame for data processing and file management, and a dedicated visualization computer with a high-resolution display, large buffers, and high-speed data paths linking components within the environment. Software support allows use of graphic techniques based upon, or in combination with, points, areas, volumes, geometric primitives, wire frames, and color at a frame rate greater than 20 frames per second.

4.1.2 Dynamic Training and Simulation

Techniques that allow operator feedback into real-time control functions that enhance realism by coordinated multisensor operator inputs. They incorporate real-time evaluation of operator/trainee performance and are being used extensively for operator training, maintenance, and repair of a wide variety of equipment.

4.1.3 Signal Processing

Digital computers or assemblies or software characterized by an architectural structure wherein the principal processing elements are optimized for the manipulation of data generally derived from external sensors and from which signals of interest can be obtained are considered to be signal processors. Information is extracted from the signals by employing equipment, algorithms, and software programs. The output of the signal extraction process is subjected to additional analysis and processing from which critical parameters and classifications are determined. The entire computational analysis must be done in real time to be of critical use.

4.1.4 Image Processing

Technology used for acquiring, transferring, analyzing, displaying and making tactical use of image data in real time or near real time. Of particular concern are those technologies related to: implementation of mobile sensors for real-time target acquisition and guidance, processing and displays of large complex data sets, archival storage of imagery data, real-time displays and three-dimensional (3-D) presentation, and data transmission and compression techniques.

4.1.6 Speech Processing Systems

Techniques for analyzing and synthesizing continuous (connected) speech exist in the form of algorithms and special semiconductor devices. Operating hardware is available and improved techniques are under development for synthesizing speech from a digital stream to provide an improved interface between computers and man.

4.2.4 Hard Real-Time Systems

Real-time processing is defined as the processing of data by a computer system to provide a required level of service, as a function of available resources, within a guaranteed response time, regardless of the load of the system, when stimulated by an external event. This definition also applies to hard real-time operating systems that provide a shared set of computer resource management services designed and optimized for supporting such time-critical computer software applications as process control systems, command and control, and flight control. Key functional attributes of real-time operating systems to protect and isolate application programs from the effects of errors in other programs sharing the resources and their ability to respond to stimuli in a guaranteed predetermined time.

4.2.5 Data Fusion

Data fusion is the technique whereby multivariate data from multiple sources are retrieved and processed as a single, unified, logical file. It is an integral part of command and control systems with distributed sensors. The range of fusion requirements increases as the need progresses from the missile seekers of an air-to-air missile to the avionics cockpit through various echelon levels to the command center. Intelligent processing is a major ingredient of all fusion requirements. A significant set of *a priori* databases are crucial to the effective functioning of the fusion process.

4.2.9 Artificial Intelligence

The field of computer science seeking to instill into machines the ability to solve problems through the careful use of knowledge in approaches such as expert reasoning, analysis, analogy (educated guessing), and learning. Vision, natural language understanding, expert systems, fuzzy logic, and neural networks are several subfields of artificial intelligence.

9.5.1 Spacecraft Structures

This item covers specifications for dimensionally stable structures for spacecraft which employ techniques for control of structural distortion, including materials designed for zero coefficient of thermal expansion designs to prevent structural outgassing in orbit and materials that provide high strength and high stiffness.

Analysis techniques are used to simulate the dynamic interaction of the structure with the spacecraft control system in lieu of full-scale testing and provide the means to define a design with the required stability characteristics for spacecraft with precision structures, including optical systems and antennas or such large flexible appendages as solar panels.

9.5.2   Nonchemical, High Specific Impulse Propulsion

This item covers low-thrust propulsion devices that can be used for spacecraft station keeping or orbit changes including electrostatic, electrothermal, or electromagnetic propulsion systems. These devices utilize electrical power to accelerate propellant gases to high-exit velocities.

9.5.4   Aircraft High-Performance Structures

This section covers dual-use technology that will improve (1) the performance of current, near-future, and advanced (including hypersonic) aircraft with structures exhibiting improved mechanical and thermal properties as well as lower weight and cost and (2) performance, durability, fatigue life, and acoustic damping of helicopters and other vertical take-off and landing (VTOL), short take-off/vertical landing (STOVL), and short take-off and landing (STOL) aircraft, whose designs involve exposure of aircraft structure to high levels of engine-generated heat and noise. This section covers dual-use technology that will increase structural safety and operational readiness and decrease maintenance and repair costs of aircraft structures. This section also covers military-use technology that will improve the efficiency of repairing battle-damaged composite structures and dual-use technology for repairing composite structures damaged in peacetime service or by maintenance mishap.

11.4.2   Kinetic Energy Projectiles

This item covers technologies for kinetic energy projectiles which can destroy or damage targets through their own kinetic energy derived from their own nonchemical sources.

11.4.4   Kinetic Energy Platform Management

This item delineates technologies necessary to ensure efficient platform management systems for kinetic-energy weapons for both space and ground-based systems.

Appendix E

System Scores

This appendix presents the assigned values of each measure of merit for each system. As described in the "Methodology" chapter, the scoring functions (fig. A-5 through fig. A-32) were used to determine the utility of the systems for each measure of merit; the utilities were then multiplied through the weighted value models (appendix C) to determine the overall system values. Figure E-7 and figure E-8 present the results of this process, which is also shown graphically in the "Results" chapter.

A keyed designator is provided to distinguish measures of merit having duplicate names (e.g., accuracy). Figure E-1 shows the Value Model with this additional keying information.

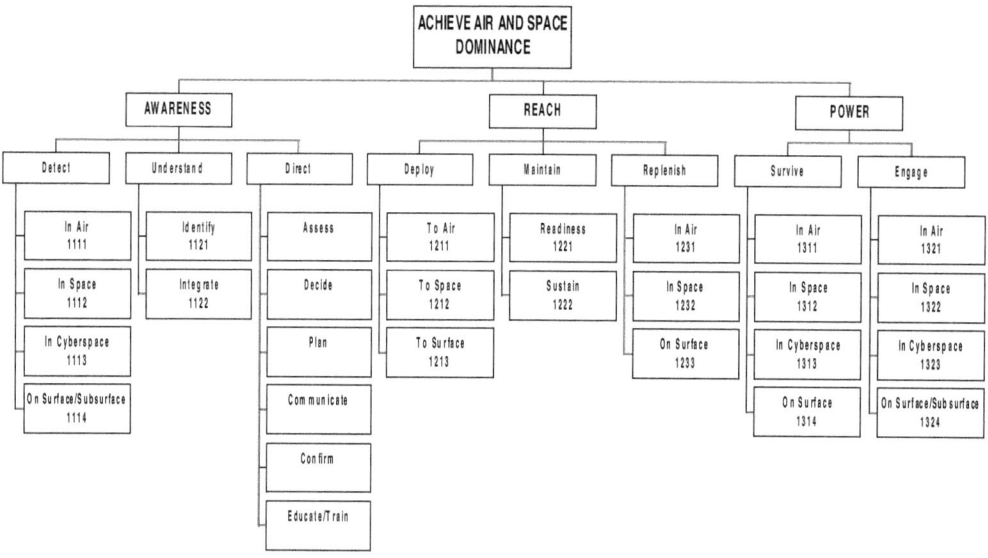

Figure E-1. Value Model - Top Level

Figure E-2. System Scoring - *Detect* Task

System Name	Coverage 1111	Timely 1111	Accurate 1111	Scope 1111	Sensor Variety 1111	Unobtrusive 1111	Resolution 1111	Coverage 1112	Timely 1112	Accurate 1112	Scope 1112	Sensor Variety 1112	Range 1112	Unobtrusive 1112	Resolution 1112	Timely 1113	Accurate 1113	Sensor Variety 1113	Unobtrusive 1113	Accurate 1114	Timely 1114	Coverage 1114	Unobtrusive 1114	Resolution 1114	Scope 1114	Sensor Variety 1114
1.1 HypS Atk Acft	Reg	1	10 - 100	90	2	100	60	0	0	>10,000	0	0	0	100	10 m	>1 day	0	0	0	1 - 10	>24	1 Reg	50	60	70	2
1.2 Fotofighter	Coun	0	0 - 1	95	3	50	90	0	0	>10,000	0	0	0	100	10 m	>1 day	0	0	0	0 - 1	>24	1 Coun	50	70	90	2
1.3 Container Aircraft	N	24	>10,000	0	2	100	0	0	>24	>10,000	0	0	0	100	10 m	>1 day	0	0	0	>10,000	>24	0 N	100	0	0	0
1.4 L-T-A Airlifter	Coun	1	10 - 100	70	0	100	50	0	>24	>10,000	0	0	0	100	10 m	>1 day	0	0	0	>10,000	>24	0 N	100	0	0	0
1.5 Supersonic Airlift	N	24	>10,000	0	0	100	0	0	>24	>10,000	0	0	0	100	10 m	>1 day	0	0	0	>10,000	>24	0 N	100	0	0	0
1.6 Stealth Airlifter	N	24	>10,000	0	0	100	0	0	>24	>10,000	0	0	0	100	10 m	>1 day	0	0	0	>10,000	>24	0 N	100	0	0	0
1.7 Glob Transport Acft	N	24	>10,000	0	0	100	0	0	>24	>10,000	0	0	0	100	10 m	>1 day	0	0	0	>10,000	>24	0 N	100	0	0	0
2.1 Strike UAV	N	24	>10,000	0	0	100	0	0	>24	>10,000	0	0	0	100	10 m	>1 day	0	0	0	>10,000	>24	0 N	100	0	0	0
2.2 Recce UAV	Coun	0	0 - 1	100	4	75	90	0	>24	>10,000	0	0	0	100	10 m	>1 sec	40	1	0	0 - 1	0	0 Coun	75	70	95	4
2.3 UCAV	Coun	0	0 - 1	100	0	75	90	0	>24	>10,000	0	0	0	100	10 m	>1 sec	40	1	0	0 - 1	0	0 Coun	75	70	95	4
2.4 PDS	N	24	>10,000	0	0	100	0	0	>24	>10,000	0	0	0	100	10 m	>1 day	0	0	0	>10,000	0	0 N	100	0	0	1
2.5 UAV Mothership	Coun	1	0 - 1	70	2	100	60	0	>24	>10,000	0	0	0	100	10 m	>1 day	0	0	0	>10,000	>24	0 N	100	0	0	0
2.6 Exfiltration Rocket	N	24	>10,000	0	0	100	0	0	>24	>10,000	0	0	0	100	10 m	>1 day	0	0	0	>10,000	>24	0 N	100	0	0	0
3.1 OMV	N	24	>10,000	0	0	100	0	0	>12	1 - 10	100	2	LEO	100	100 meter	>1 day	0	0	0	>10,000	>24	0 N	100	0	0	0
3.2 OCV	N	24	>10,000	0	0	100	0	0	>12	0 - 1	100	2	LEO	100	75 cm	>1 day	0	0	0	>10,000	>24	0 N	100	0	0	0
3.3 SB	N	24	>10,000	0	0	100	0	20	20	0 - 1	100	2	Geo	100	100 meter	>1 day	0	0	0	>1 - 10	>24	1 Reg	100	40	70	2
4.1 Piloted SSTO TAV	Reg	1	1 - 10	95	3	100	60	50	0	0 - 1	90	2	Geo	100	100 meter	>1 day	0	0	0	>1 - 10	>24	1 Reg	100	40	70	2
4.2 Uninhab AL TAV	Reg	1	1 - 10	95	3	100	60	50	0	0 - 1	0	0	0	100	10 m	>1 day	0	0	0	>10,000	>24	0 N	100	0	0	0
5.1 AYM	N	24	>10,000	0	5	100	90	0	0	>10,000	0	0	0	100	10 m	>1 day	0	0	0	>10,000	>24	0 N	100	0	0	0
5.2 AAAM	N	24	0 - 1	90	0	100	0	0	>24	>10,000	0	0	0	100	10 m	>1 day	0	0	0	>10,000	>24	0 N	100	0	0	0
5.3 Airborne HPMW	N	24	>10,000	0	0	100	0	0	>24	>10,000	0	0	0	100	10 m	>1 day	0	0	0	>10,000	>24	0 N	100	0	0	0
5.4 SHM	N	24	>10,000	0	0	100	0	0	>24	>10,000	0	0	0	100	10 m	>1 day	0	0	0	>1 - 10	>24	1 Coun	50	40	90	2
5.5 Attack Microbots	N	10	0 - 1	50	2	100	20	0	0	0 - 1	0	0	0	100	10 m	>1 sec	20	1	0	0 - 1	0	0 N	20	20	70	0
5.6 Abn Holo Projector	N	24	>10,000	0	0	100	0	0	>24	>10,000	0	0	0	100	10 m	>1 day	0	0	0	>10,000	>24	0 N	100	0	0	0
5.7 Hybrid HEL Sys	N	24	>10,000	0	0	100	0	15	6	0 - 1	80	2	moon	0 cm	0 cm	>1 day	0	0	0	>1 - 10	>24	1 Reg	100	40	70	2
6.1 GLASS	Reg	1	1 - 10	95	3	100	80	50	0	0 - 1	90	2	Geo	100 meter	100 meter	>1 day	0	0	0	>1 - 10	>24	1 Reg	100	40	70	2
6.2 Space KEW	Glob	1	1 - 10	75	2	65	70	100	0	>10,000	100	2	moon	65 cm	65 cm	>1 day	0	0	0	>10,000	>24	0 Glob	65	70	50	2
6.3 Space HPMW	Glob	0	0 - 1	75	2	65	70	100	0	10000	100	2	moon	65 cm	65 cm	>1 day	0	0	0	>10,000	>24	0 Glob	65	70	50	2
6.4 Space HEL	N	24	>10,000	0	0	100	0	0	>24	0 - 1	0	0	0	100	10 m	>1 day	0	0	0	>10,000	>24	0 N	100	0	0	0
6.5 SolarHEL	N	24	>10,000	0	0	100	0	0	>24	0 - 1	0	0	0	100	10 m	>1 day	0	0	0	>10,000	>24	0 N	100	0	0	0
6.6 SEOW	N	24	>10,000	0	0	100	0	0	>24	>10,000	0	0	0	100	10 m	>1 day	0	0	0	>10,000	>24	0 N	100	0	0	0
6.7 AMS	N	24	>10,000	0	0	100	0	0	>24	>10,000	0	0	0	100	10 m	>1 day	0	0	0	>10,000	>24	0 N	100	0	0	0
7.1 SLT	N	24	>10,000	0	0	100	0	0	>24	>10,000	0	0	0	100	10 m	>1 day	0	0	0	>10,000	>24	0 N	100	0	0	0
7.2 PDA	N	24	>10,000	0	0	100	0	0	>24	>10,000	0	0	0	100	10 m	>1 day	0	0	0	>10,000	>24	0 N	100	0	0	0
7.3 Vis. Interactive Ctr	N	24	>10,000	0	0	100	0	0	>24	>10,000	0	0	0	100	10 m	>1 day	0	0	0	>10,000	>24	0 N	100	0	0	0
8.1 GIMS	Glob	0	0 - 1	100	6	20	100	100	0	0 - 1	100	4	moon	20 cm	20 cm	>1 sec	80	4	50	0 - 1	0	0 Glob	20	100	100	6
8.2 GSRT	Glob	0	0 - 1	100	0	65	80	100	0	0 - 1	100	4	moon	65 cm	65 cm	>1 sec	40	3	0	0 - 1	0	0 Glob	65	80	80	6
8.3 Sensor Microbots	N	1	0 - 1	50	6	65	20	0	0	0 - 1	0	0	0	100	10 m	>1 sec	0	0	0	0 - 1	>1	1 N	20	20	70	6
8.4 ML Sensor Sys	Coun	1	0 - 1	90	3	100	70	0	>24	>10,000	95	2	moon	100	10 m	>1 sec	60	0	0	>10,000	>24	1 N	100	80	90	2
8.5 ADS	N	24	>10,000	0	0	100	0	0	>24	>0,000	0	0	0	100	10 m	>1 day	0	0	0	>10,000	>24	0 N	100	0	0	0
9.1 MARS	N	24	>10,000	0	0	100	0	0	>24	>0,000	0	0	0	100	10 m	>1 day	0	0	0	>10,000	>24	0 N	100	0	0	0
9.2 WAMS	N	6	1 - 10	30	1	100	0	0	>24	>10,000	0	0	0	100	10 m	>1 day	0	0	0	>10,000	>24	0 N	100	0	0	0
9.3 Sanctuary Base	N	0	0 - 1	100	6	20	90	0	>24	>10,000	0	0	0	100	10 m	>1 day	0	0	0	>10,000	0	0 N	20	90	100	6

System Name	Accurate 1121	Timely 1121	Traceable	Battlespace View	Timely 1122	Correlation 1122	Own Force Status	Other's Status	Environment	Decision Basis	Quality	Effective	Efficient	Security	Tailored (Sys-Hmn)	Interaction(Hmn-Hmn)	Capacity	Interoperable	BDA	Feedback	Doctrine	TechOps Knowledge	TechOps Skills	Critical/Creative T
1.1 HypS Atk Acft	90	JIT	80	10	JIT	0	50	50	50	0	0	0	0	0	0	N	0	0	0	24	N	0	0	N
1.2 Fotofighter	90	JIT	80	10	JIT	0	50	50	50	0	0	0	0	0	0	N	0	0	90	0	N	0	0	N
1.3 Container Aircraft	0	TL	0	50	JIT	25	75	60	75	50	50	50	50	70	50	VTC	2	4	0	24	N	0	0	N
1.4 L-T-A Airlifter	0	TL	0	5	TL	0	50	50	50	0	0	0	0	0	0	N	0	0	0	24	N	0	0	N
1.5 Supersonic Airlift	0	TL	0	0	TL	0	50	50	50	0	0	0	0	0	0	N	0	0	0	24	N	0	0	N
1.6 Stealth Airlifter	0	TL	0	0	TL	0	50	50	50	0	0	0	0	0	0	N	0	0	0	24	N	0	0	N
1.7 Glob Transport Acft	0	TL	0	0	TL	0	50	50	50	0	0	0	0	0	0	N	0	0	0	24	N	0	0	N
2.1 Strike UAV	95	IA	90	10	JIT	10	50	50	50	0	0	0	0	0	0	N	0	0	90	0.5	N	0	0	N
2.2 Recce UAV	95	IA	90	10	JIT	10	50	50	50	0	0	0	0	0	0	N	0	0	90	0	N	0	0	N
2.3 UCAV	0	TL	0	0	TL	0	50	50	50	0	0	0	0	0	0	N	0	0	0	24	N	0	0	N
2.4 PDS	0	TL	0	10	JIT	10	50	50	50	0	0	0	0	0	0	N	0	0	0	24	N	0	0	N
2.5 UAV Mothership	0	TL	0	0	TL	0	50	50	50	0	0	0	0	0	0	N	0	0	0	24	N	0	0	N
2.6 Exfiltration Rocket	0	TL	0	0	TL	0	50	50	50	0	0	0	0	0	0	N	0	0	0	24	N	0	0	N
3.1 OMV	90	JIT	90	10	JIT	10	50	50	50	0	0	0	0	0	0	N	0	0	90	0	N	0	0	N
3.2 OCV	90	JIT	90	10	JIT	10	50	50	50	0	0	0	0	0	0	N	0	0	90	1	N	0	0	N
3.3 SB	80	JIT	80	10	JIT	0	50	50	50	0	0	0	0	0	0	N	0	0	80	1	N	0	0	N
4.1 Piloted SSTO TAV	80	JIT	90	10	JIT	0	50	50	50	0	0	0	0	0	0	N	0	0	80	1	N	0	0	N
4.2 Uninhab AL TAV	0	TL	0	0	TL	0	50	50	50	0	0	0	0	0	0	N	0	0	0	24	N	0	0	N
5.1 AYM	0	TL	0	0	TL	0	50	50	50	0	0	0	0	0	0	N	0	0	0	24	N	0	0	N
5.2 AAAM	0	TL	0	0	TL	0	50	50	50	0	0	0	0	0	0	N	0	0	0	24	N	0	0	N
5.3 Airborne HPMW	0	TL	0	0	TL	0	50	50	50	0	0	0	0	0	0	N	0	0	0	24	N	0	0	N
5.4 SHM	0	TL	0	0	TL	0	50	50	50	0	0	0	0	0	0	N	0	0	0	24	N	0	0	N
5.5 Attack Microbots	0	TL	0	0	JIT	0	50	50	50	0	0	0	0	0	0	N	0	0	0	24	N	0	0	N
5.6 Abn Holo Projector	0	TL	0	0	TL	0	50	50	50	0	0	0	0	0	0	N	0	0	0	1	N	0	0	N
5.7 Hybrid HEL Sys	0	TL	0	0	TL	0	50	50	50	0	0	0	0	0	0	N	0	0	15	0	N	0	0	N
6.1 GLASS	80	JIT	80	10	JIT	0	50	50	50	0	0	0	0	0	0	N	0	0	80	1	N	0	0	N
6.2 Space KEW	0	TL	0	0	TL	0	50	50	50	0	0	0	0	0	0	N	0	0	0	24	N	0	0	N
6.3 Space HPMW	0	IA	90	0	IA	0	50	50	50	0	0	0	0	0	0	N	0	0	60	0	N	0	0	N
6.4 Space HEL	90	IA	90	0	IA	0	50	50	50	0	0	0	0	0	0	N	0	0	60	0	N	0	0	N
6.5 SolarHEL	0	TL	0	0	TL	0	50	50	50	0	0	0	0	0	0	N	0	0	0	24	N	0	0	N
6.6 SEOW	0	TL	0	0	TL	0	50	50	50	0	0	0	0	0	0	N	0	0	0	24	N	0	0	N
6.7 AMS	0	TL	0	20	TL	20	50	50	50	25	0	20	20	0	0	N	0	0	0	24	N	0	0	N
7.1 SLT	0	TL	0	50	IA	50	70	70	60	50	0	50	50	95	100	Voice	2	7	50	0	Comp Eval	20 70	50 90	Comp Synth
7.2 PDA	0	TL	0	100	IA	100	95	90	90	90	80	100	100	95	100	VTC	0	7	0	24	Comp Eval	70 80	90 95	Comp Synth
7.3 Vis. Interactive Ctr	0	TL	0	100	IA	100	100	100	100	100	100	100	100	95	100	VR	20	7	100	0	Eval	80	95	Synth
8.1 GIMS	99	IA	90	70	IA	70	100	80	80	80	0	0	80	90	50	VR	15	7	95	0	N	0	0	N
8.3 GSRT	95	IA	90	0	TL	0	50	50	50	0	0	0	0	0	0	N	0	0	50	1	N	0	0	N
8.4 Sensor Microbots	0	TL	0	0	TL	0	50	50	50	0	0	0	0	0	0	N	0	0	50	0	N	0	0	N
8.5 ML Sensor Sys	95	IA	100	0	JIT	0	50	50	50	0	0	0	0	0	0	N	0	0	100	24	N	0	0	N
9.1 ADS	0	TL	0	0	TL	0	50	50	50	0	0	0	0	0	0	N	0	0	0	24	N	0	0	N
9.2 WAMS	0	TL	0	0	JIT	0	50	50	80	0	0	0	0	0	0	N	0	0	0	24	N	0	0	N
9.3 Sanctuary Base	100	JIT	100	100	JIT	50	50	50	50	90	80	100	100	95	100	VR	5	7	90	0	N	0	0	N

Figure E-3. System Scoring - *Understand/Direct* Tasks

System Name	Payload Weight 1211	Payload Volume 1211	Range 1211	Responsive 1211	Multirole 1211	Payload Weight 1212	Payload Volume 1212	Responsive 1212	Reusable 1212	Specific Impulse	Perf.A, Surf-Space	Perf.B, Space-Space	Payload Weight 1213	Payload Volume 1213	Range 1213	Forward Basing Reqt	Available 1221	Footprint 1222	Supportable 1221	Pipeline 1222	Responsive 1213	Multirole 1213	Timely 1231	Accurate 1231	Transfer 1231	Multicommodity 1231	Transfer Rate 1231	Timely 1232	Accurate 1232	Transfer 1232	Reusable 1232	Transfer Rate 1232	Timely 1233	Accurate 1233	Transfer 1233
1.1 HypS Atk Acft	50	20	10	0.5	1	0	0	30	0	0	0	0	0	0	0	Huge/N.A.	80	S	80	Depot	10	0	24	80	0	0	60	7	<100%	0	0	W	10	Thea	0
1.2 Fotofighter	0	0	0	10	0	0	0	30	0	0	0	0	0	0	0	Huge/N.A.	80	M	70	Depot	10	0	24	80	0	0	60	7	<100%	0	0	W	10	Thea	0
1.3 Container Aircraft	0	100	0	10	3	20	15	30	0	0	0	0	200	50	10	Bare Base	80	M	80	Depot	2	3	24	80	97	4	60	7	<100%	0	0	W	2	FOB	100
1.4 L-T-A Airlifter	250	0	12.5	7	3	20	15	30	0	0	0	0	200	100	12.5	Bare Base	85	M	80	Depot	5	3	12	100	0	0	60	7	<100%	0	0	W	7	FOB	100
1.5 Supersonic Airlift	0	0	12.5	10	0	5	2	30	0	0	0	0	50	50	12.5	Full Base	75	M	80	Depot	0.5	0	3	100	0	0	60	7	<100%	0	0	W	0.5	FOB	100
1.6 Stealth Airlifter	0	0	10	10	1	3	0.65	30	0	0	0	0	10	12.5	5	Full Base	80	S	70	Depot	1	2	24	80	0	1	60	7	<100%	0	0	W	1	Unit	100
1.7 Glob Transport Ac	0	0	0	10	0	0	0	30	0	0	0	0	250	75	12	N	90	M	90	Mx sqdn	2	3	3	100	97	1	5	7	<100%	0	0	W	10	Thea	0
2.1 Strike UAV	0	0	0	10	0	0	0	30	0	0	0	0	0	0	0	Huge/N.A.	90	S	90	Mx sqdn	10	0	24	80	0	0	60	7	<100%	0	0	W	10	Thea	0
2.2 Recce UAV	0	0	0	10	0	0	0	30	0	0	0	0	0	0	0	Huge/N.A.	90	S	90	Mx sqdn	10	0	24	80	0	0	60	7	<100%	0	0	W	10	Thea	0
2.3 UCAV	0	0	0	10	0	0	0	30	0	0	0	0	5	0	0	Huge/N.A.	90	S	95	Mx sqdn	10	2	24	80	0	0	60	7	<100%	0	0	W	0.5	Sys	95
2.4 PDS	0	0	0	10	0	0	0	30	0	0	0	0	1	1	0.5	Huge/N.A.	80	M	90	Mx sqdn	0.5	0	24	80	0	0	60	7	<100%	0	0	W	10	Thea	0
2.6 UAV Mothership	50	25	5	10	0.5	1	0	30	0	0	0	0	0	0	0	Huge/N.A.	80	M	25	Depot	0	1	24	80	0	5	60	7	1	0	0	W	10	Thea	0
2.6 Exfiltration Rocket	0	0	0	10	0	0	0	30	100	1,000	0	5	0	0	0	Huge/N.A.	80	M	50	Depot	10	0	24	80	0	0	60	2	1	100	100	W	10	Thea	0
3.1 OMV	0	0	0	10	0	0	0	30	100	1,000	0	5	0	0	0	Huge/N.A.	80	M	50	Depot	10	0	24	80	0	0	60	7	1	100	100	H	10	Thea	0
3.2 OCV	0	0	0	10	0	0	0	30	0	0	0	0	0	0	0	Huge/N.A.	80	S	50	Depot	10	0	24	80	0	0	60	7	<100%	0	0	H	10	Thea	0
3.3 SB	0	0	0	10	0	5	2	30	0	0	45	0	0	0	0	Huge/N.A.	80	S	80	Depot	10	0	24	80	0	0	60	7	<100%	0	0	H	10	Thea	0
4.1 Piloted SSTO TAV	0	0	0	10	0	5	2	0.5	100	1,000	45	0	0	0	0	Huge/N.A.	70	S	95	5 vans	10	0	24	80	0	0	60	0.5	1	100	100	H	10	Thea	0
4.2 Uninhab AL TAV	0	0	0	10	0	3	0	0.5	100	1,000	45	0	0	0	0	Huge/N.A.	95	S	95	5 vans	10	0	24	80	0	0	60	0.5	1	100	100	H	10	Thea	0
5.1 AYM	0	0	0	10	0	0	0	30	0	0	0	0	0	0	0	Huge/N.A.	95	S	95	5 vans	10	0	24	80	0	0	60	7	<100%	0	0	W	10	Thea	0
5.2 AAAM	0	0	0	10	0	0	0	30	0	0	0	0	0	0	0	Huge/N.A.	95	S	95	Mx sqdn	10	0	24	80	0	0	60	7	<100%	0	0	W	10	Thea	0
5.3 Airborne H>MW	0	0	0	10	0	0	0	30	0	0	0	0	0	0	0	Huge/N.A.	90	S	90	5 vans	10	0	24	80	0	0	60	7	<100%	0	0	W	10	Thea	0
5.4 SHM	0	0	0	10	0	0	0	30	0	0	0	0	0	0	0	Huge/N.A.	100	M	70	10*Depot	10	0	24	80	0	0	60	7	<100%	0	0	M	10	Thea	0
5.5 Attack Microbots	0	0	0	10	0	0	0	30	0	0	0	0	0	0	0	Huge/N.A.	100	M	50	10*Depot	10	0	24	80	0	0	60	7	<100%	0	0	H	10	Thea	0
5.6 Abn Holo Projecto	0	0	0	10	0	5	2	0.5	100	1,000	45	0	0	0	0	Huge/N.A.	100	M	50	10*Depot	10	0	24	80	0	0	60	0.1	1	95	100	H	10	Thea	0
5.7 Hybrid HEL Sys	0	0	0	10	0	0	0	30	0	0	0	0	0	0	0	Huge/N.A.	100	M	70	10*Depot	10	0	24	80	0	0	60	0.5	1	100	100	M	10	Thea	0
6.1 GLASS	0	0	0	10	0	0	0	30	0	0	0	0	0	0	0	Huge/N.A.	100	S	50	10*Depot	10	0	24	80	0	0	60	7	<100%	0	0	W	10	Thea	0
6.2 Space KEW	0	0	0	10	0	0	0	30	0	0	0	0	0	0	0	Huge/N.A.	100	M	50	Depot	10	0	24	80	0	0	60	7	<100%	0	0	W	10	Thea	0
6.3 Space HPMW	0	0	0	10	0	0	0	30	0	0	0	0	0	0	0	Huge/N.A.	100	S	60	Depot	10	0	24	80	0	0	60	7	<100%	0	0	W	10	Thea	0
6.4 Space HEL	0	0	0	10	0	0	0	30	0	0	0	0	0	0	0	Huge/N.A.	95	N	40	Mx sqdn	10	0	24	80	0	0	60	7	<100%	0	0	W	10	Thea	0
6.5 Solar HEL	0	0	0	10	0	0	0	30	0	0	0	0	0	0	0	Huge/N.A.	95	N	95	10*Depot	10	0	24	80	0	0	60	7	<100%	0	0	W	10	Thea	0
6.6 SEOW	0	0	0	10	0	0	0	30	0	0	0	0	0	0	0	Huge/N.A.	100	M	50	10*Depot	10	0	24	80	0	0	60	7	<100%	0	0	W	10	Thea	0
6.7 AMS	0	0	0	10	0	0	0	30	0	0	0	0	0	0	0	Huge/N.A.	95	M	0	Mx sqdn	10	0	24	80	0	0	60	7	<100%	0	0	W	10	Thea	0
7.1 SLT	0	0	0	10	0	0	0	30	0	0	0	0	0	0	0	Huge/N.A.	100	S	60	Depot	10	0	24	80	0	0	60	7	<100%	0	0	W	10	Thea	0
7.2 PDA	0	0	0	10	0	0	0	30	0	0	0	0	0	0	0	Huge/N.A.	95	N	40	Mx sqdn	10	0	24	80	0	0	60	7	<100%	0	0	W	10	Thea	0
7.3 Vis. Interactive Ctr	0	0	0	10	0	0	0	30	0	0	0	0	0	0	0	Huge/N.A.	95	N	95	10*Depot	10	0	24	80	0	0	60	7	<100%	0	0	W	10	Thea	0
8.1 GIMS	0	0	0	10	0	0	0	30	0	0	0	0	0	0	0	Huge/N.A.	100	M	50	10*Depot	10	0	24	80	0	0	60	7	<100%	0	0	W	10	Thea	0
8.2 GSRT	0	0	0	10	0	0	0	30	0	0	0	0	0	0	0	Huge/N.A.	95	M	0	Mx sqdn	10	0	24	80	0	0	60	7	<100%	0	0	W	10	Thea	0
8.3 Sensor Microbots	0	0	0	10	0	0	0	30	0	0	0	0	0	0	0	Huge/N.A.	80	S	90	5 vans	10	0	24	80	0	0	60	7	<100%	0	0	W	10	Thea	0
8.4 ML Sensor Sys	0	0	0	10	0	0	0	30	0	0	0	0	0	0	0	Huge/N.A.	100	M	60	Depot	10	0	24	80	0	0	60	7	<100%	0	0	W	10	Thea	0
8.5 ADS	0	0	0	10	0	0	0	30	0	0	0	0	0	0	0	Huge/N.A.	95	N	##	10*Depot	10	0	24	80	0	0	60	7	<100%	0	0	W	10	Thea	0
9.1 MARS	0	0	0	10	0	0	0	30	0	0	0	0	0	0	0	Huge/N.A.	100	L	90	Mx sqdn	10	0	24	80	0	0	60	7	<100%	0	0	W	10	Thea	0
9.2 WAMS	0	0	0	10	0	0	0	30	0	0	0	0	0	0	0	Huge/N.A.	95	N	##	10*Depot	10	0	24	80	0	0	60	7	<100%	0	0	W	10	Thea	0
9.3 Sanctuary Base	0	0	0	10	0	0	0	30	0	0	0	0	0	0	0	Huge/N.A.	100	N	0	Depot	10	0	24	80	0	0	60	7	<100%	0	0	W	0	Sys	100

Figure E-4. System Scoring - *Deploy/Maintain/Replenish* Tasks

	System Name	Detectable 1311	Countermeasures 1311	Vulnerable 1311	Recoverable 1311	Detectable 1312	Countermeasures 1312	Vulnerable 1312	Recoverable 1312	Detectable 1313	Countermeasures 1313	Vulnerable 1313	Recoverable 1313	Detectable 1314	Countermeasures 1314	Vulnerable 1314	Recoverable 1314
1.1	HypS Atk Acft	0.8	0.1	0.5	24	1	1	1	24	1	0	1	24	1	1	1	24
1.2	Fotofighter	0.5	0.1	0.5	24	1	1	1	24	1	0	1	24	1	1	1	24
1.3	Container Aircraft	1	1	1	24	1	1	1	24	0.8	1	0.6	3	0.9	0.5	0.5	24
1.4	L-T-A Airlifter	1	1	0.6	24	1	1	1	24	1	0	1	24	1	1	1	24
1.5	Supersonic Airlift	0.9	0.5	0.8	24	1	1	1	24	1	0	1	24	1	1	1	24
1.6	Stealth Airlifter	0.1	0.3	0.5	24	1	1	1	24	1	0	1	24	1	1	1	24
1.7	Glob Transport Acf	1	1	1	24	1	1	1	24	1	0	1	24	1	1	1	24
2.1	Strike UAV	0.3	0.3	0.5	24	1	1	1	24	1	0	1	24	1	1	1	24
2.2	Recce UAV	0.5	0.4	0.5	12	1	1	1	24	1	0	1	24	1	1	1	24
2.3	UCAV	0.5	0.4	0.5	12	1	1	1	24	1	0	1	24	1	1	1	24
2.4	PDS	0.7	0.7	0.5	12	1	1	1	24	1	0	1	24	1	1	1	24
2.5	UAV Mothership	1	0.9	0.6	24	1	1	1	24	1	0	1	24	1	1	1	24
2.6	Exfiltration Rocket	1	0.8	1	24	1	0.7	0.9	24	1	0	1	24	0.2	1	1	24
3.1	OMV	1	1	1	24	1	0.4	0.9	24	1	0	1	24	1	1	1	24
3.2	OCV	1	1	1	24	1	0.4	0.5	24	1	0	1	24	1	1	1	24
3.3	SB	1	1	1	24	0.7	0.4	0.9	24	1	0	1	24	1	1	1	24
4.1	Piloted SSTO TAV	1	1	1	24	1	1	1	24	1	0	1	24	1	1	1	24
4.2	Uninhab AL TAV	1	1	1	24	1	1	1	24	1	0	1	24	1	1	1	24
5.1	AYM	1	1	1	24	1	1	1	24	1	0	1	24	1	1	1	24
5.2	AAAM	1	1	1	24	1	1	1	24	1	0	1	24	1	1	1	24
5.3	Airborne HPMW	1	1	1	24	1	1	1	24	1	0	1	24	1	1	1	24
5.4	SHM	0.6	0.1	1	24	1	0.4	0.1	12	0.1	0	0.1	24	0.1	0.1	0.1	24
5.5	Attack Microbots	1	0.2	1	24	1	0.7	0.1	6	1	0	1	24	1	1	1	12
5.6	Abn Holo Projector	0.5	0.1	0.1	12	1	0.4	0.1	12	1	0	1	24	1	0.5	0.3	24
5.7	Hybrid HEL Sys	1	0.5	1	24	1	0.7	0.1	12	1	0	1	24	1	0.5	1	24
6.1	GLASS	1	1	1	24	1	1	0.9	24	1	0	1	24	1	1	1	24
6.2	Space KEW	1	1	1	24	1	0.4	0.1	12	1	0	1	24	1	1	1	24
6.3	Space HPMW	1	1	1	24	1	0.7	0.1	6	1	0	1	24	1	1	1	24
6.4	Space HEL	1	1	1	24	1	0.4	0.1	12	1	0	1	24	1	1	1	24
6.5	SolarHEL	1	1	1	24	1	0.7	0.9	24	1	0	1	24	1	1	1	24
6.6	SEOW	1	1	1	24	1	1	1	24	1	0	1	24	1	1	1	24
6.7	AMS	1	1	1	24	1	1	1	24	1	0	1	24	1	1	1	24
7.1	SLT	1	1	1	24	1	1	1	24	1	0	1	24	1	1	1	24
7.2	PDA	1	1	1	24	1	1	1	24	1	0	1	24	1	1	1	24
7.3	Vis. Interactive Ctr	1	1	1	24	1	1	1	24	1	0	1	24	1	1	1	24
8.1	GIMS	0.1	0.1	0.1	6	1	0.7	0.1	6	0.1	5	0.3	1	0.2	0.5	0.3	6
8.2	GSRT	1	1	1	24	1	0.7	0.1	6	0.5	3	0.3	3	1	0.1	0.1	24
8.3	Sensor Microbots	0.1	0.1	0.1	24	1	1	1	24	0.1	0	0.1	24	0.1	1	0.1	24
8.4	ML Sensor Sys	1	1	1	24	1	1	1	24	1	0	1	24	1	1	1	24
8.5	ADS	1	1	1	24	1	1	1	24	1	0	1	24	1	1	1	24
9.1	MARS	1	1	1	24	1	1	1	24	1	0	1	24	1	1	1	24
9.2	WAMS	0.8	0.5	1	24	1	1	1	24	1	0	0.6	3	0.5	1	1	24
9.3	Sanctuary Base	1	1	1	24	1	1	1	24	0.8	1	0.6	3	0.5	0.01	0.2	6

Figure E-5. System Scoring - *Survive* Task

System Name	Precise 1321	Timely 1321	Range 1321	Desired Lethal 1321	Reusable P-form 1321	WeaponCapacity 1321	Multirole 1321	Unobtrusive 1321	Scope 1321	Timely 1322	Desired Lethal 1322	Reusable P-form 1322	WeaponCapacity 1322	Unobtrusive 1322	Scope 1322	Range 1322	Multirole 1322	Precise 1323	Timely 1323	Desired Lethal 1323	Multirole 1323	Unobtrusive 1323	Precise 1324	Timely 1324	Engagement/v 1324	Desired Lethal 1324	Reusable P-form 1324	WeaponCapacity 1324	Multirole 1324	Unobtrusive 1324	Scope 1324
1.1 HypS Atk Acft	1 - 10	1	0	0.2	100	10	1	100	80	24	0	0	0	100	0	0	2	FR	60	0	0	100	0 - 1	1	1	0.8	100	10	2	100	80
1.2 Fotofighter	0 - 1	24	0	0.9	1,000	100	3	80	80	24	0	0	0	100	0	0	0	FR	60	0	0	100	0 - 1	24	0	0.5	1,000	100	3	80	70
1.3 Container Aircraft	> 1,000	24	0	0	0	0	0	100	0	24	0	0	0	100	0	0	0	FR	60	0	0	100	> 1,000	24	0	0	0	0	0	100	0
1.4 L-T-A Airlifter	> 1,000	24	0	0	0	0	0	100	0	24	0	0	0	100	0	0	0	FR	60	0	0	100	> 1,000	24	0	0	0	0	0	100	0
1.5 Supersonic Airlift	> 1,000	24	0	0	0	0	0	100	0	24	0	0	0	100	0	0	0	FR	60	0	0	100	> 1,000	24	0	0	0	0	0	100	0
1.6 Stealth Airlifter	> 1,000	24	0	0	0	0	0	100	0	24	0	0	0	100	0	0	0	FR	60	0	0	100	> 1,000	24	0	0	0	0	0	100	0
1.7 Glob Transport Acft	> 1,000	24	0	0	0	0	0	100	0	24	0	0	0	100	0	0	0	FR	60	0	0	100	> 1,000	24	0	0	0	0	2	30	80
2.1 Strike UAV	1 - 10	1	0	0.6	100	10	1	30	80	24	0	0	0	100	0	0	0	FR	60	0.8	1	100	0 - 1	1	0.5	0.8	100	10	0	100	80
2.2 Recce UAV	> 1,000	24	0	0	0	0	0	100	0	24	0	0	0	100	0	0	0	FR	60	0	0	100	> 1,000	24	0	0	0	0	2	100	0
2.3 UCAV	1 - 10	1	0	0.6	100	10	2	100	80	24	0	0	0	100	0	0	0	FR	60	0	0	100	0 - 1	1	0.5	0.8	100	10	2	100	80
2.4 PDS	> 1,000	24	0	0	0	0	0	100	0	24	0	0	0	100	0	0	0	ED	0	0.8	1	100	> 1,000	24	0	0	0	0	0	100	0
2.5 UAV Mothership	> 1,000	24	0	0	0	0	0	100	0	24	0	0	0	100	0	0	0	FR	60	0	0	100	> 1,000	24	0	0	0	0	0	100	0
2.6 Exfiltration Rocket	> 1,000	24	0	0	0	0	0	100	0	24	0	0	0	100	0	0	0	FR	60	0	0	100	> 1,000	24	0	0	0	0	0	100	0
3.1 OMV	> 1,000	24	0	0	0	0	0	100	0	24	0	0	0	100	0	0	0	FR	60	0	0	100	> 1,000	24	0	0	0	0	0	100	0
3.2 OCV	> 1,000	24	0	0	0	0	0	100	0	0	0.8	100	1,000	100	100	5	2	FR	60	0	0	100	> 1,000	24	0	0	0	0	0	100	0
3.3 SB	> 1,000	24	0	0	0	0	0	100	0	0	0.8	10	10	100	100	5	2	FR	60	0	0	100	> 1,000	24	0	0	0	0	0	100	0
4.1 Piloted SSTO TAV	> 1,000	24	0	0	0	0	0	100	0	1	0.95	100	10	100	100	12.5	1	FR	60	0	0	100	> 1,000	24	0	0	0	0	0	100	0
4.2 Uninhab AL TAV	> 1,000	24	0	0	0	0	0	100	0	1	0.95	100	10	100	100	12.5	1	FR	60	0	0	100	0 - 1	1	0	0.9	1	1	2	20	100
5.1 AYM	> 1,000	24	0	0	1	1	0	100	0	24	0	0	0	100	0	0	0	FR	60	0	0	100	> 1,000	24	0	0	0	0	0	100	0
5.2 AAM	1 - 10	0	0	0.9	1	1	1	100	100	24	0	0	0	100	0	0	0	CD	0	0.7	2	100	10 - 100	1	0	0.5	1,000	100	3	100	90
5.3 Airborne HPMW	10 - 100	1	0	0.7	1,000	100	3	80	100	24	0	0	0	100	0	0	0	FR	60	0.9	3	10	0 - 1	1	1.5	0.9	1	1	1	40	100
5.4 SHM	> 1,000	24	0	0	0	0	0	100	0	24	0	0	0	100	0	0	0	ED	0	0	0	100	0 - 1	1	0	0.5	1	1	2	10	70
5.5 Attack Microbots	0 - 1	1	0	0.2	0	1	2	10	50	24	0	0	0	100	0	0	0	FR	60	0	0	100	0 - 1	0	0	0.5	1	1	2	10	75
5.6 Abn Holo Projector	> 1,000	24	0	0	0	0	0	100	100	24	0	0	0	100	0	0	0	FR	60	0	0	100	> 1,000	0	0	0.5	1,000	100	1	50	50
5.7 Hybrid HEL Sys	0 - 1	0	12	0.8	1,000	1,000	3	100	75	0	0.95	1,000	1,000	100	100	25	3	FR	60	0	0	100	0 - 1	0	12	0.7	1,000	1,000	3	80	100
6.1 GLASS	0 - 1	0	12	0.9	1,000	1,000	3	80	100	0	0.95	1,000	1,000	80	100	25	3	FR	60	0	0	100	0 - 1	0	12	0.95	1,000	1,000	3	100	100
6.2 Space KEW	0 - 1	0	12	0.9	1,000	1,000	1	50	90	0	0.95	1,000	1,000	50	100	25	1	FR	60	0	0	100	0 - 1	0	12	0.95	1,000	1,000	1	50	100
6.3 Space HPMW	10 - 100	0	12	0.6	1,000	1,000	3	100	100	0	0.95	1,000	1,000	100	100	25	3	CD	0	0.5	2	100	100 - 1,000	0	12	0.4	1,000	1,000	3	80	80
6.4 Space HEL	0 - 1	0	12	0.8	1,000	1,000	3	100	75	0	0.95	1,000	1,000	100	100	25	3	FR	60	0	0	100	0 - 1	0	12	0.7	1,000	1,000	3	100	50
6.5 Solar HEL	0 - 1	0	12	0.8	1,000	100	3	100	100	0	0.4	0	100	100	100	25	3	FR	60	0	0	100	100 - 1,000	0	12	0.2	1,000	100	1	100	50
6.6 SEOW	> 1,000	24	0	0	0	0	0	100	0	0	0.1	1	10	50	50	0	0	FR	60	0	0	100	> 1,000	24	0	0	0	0	0	100	0
6.7 AMS	> 1,000	24	0	0	0	0	0	100	0	24	0	0	0	100	0	0	0	FR	60	0	0	100	> 1,000	24	0	0	0	0	0	100	0
7.1 SLT	> 1,000	24	0	0	0	0	0	100	0	24	0	0	0	100	0	0	0	FR	60	0	0	100	> 1,000	24	0	0	0	0	0	100	0
7.2 PDA	> 1,000	24	0	0	0	0	0	100	0	24	0	0	0	100	0	0	0	FR	60	0	0	100	> 1,000	24	0	0	0	0	0	100	0
7.3 Vis. Interactive Ctr	> 1,000	24	0	0	0	0	0	100	0	24	0	0	0	100	0	0	0	FR	60	0	0	100	> 1,000	24	0	0	0	0	0	100	0
8.1 GIMS	> 1,000	24	0	0	0	0	0	100	0	24	0	0	0	100	0	0	0	FR	60	0	0	100	> 1,000	24	0	0	0	0	0	100	0
8.2 GSRT	> 1,000	24	0	0	0	0	0	100	0	24	0	0	0	100	0	0	0	FR	60	0	0	100	> 1,000	24	0	0	0	0	0	100	0
8.3 Sensor Microbots	> 1,000	24	0	0	0	0	0	100	0	24	0	0	0	100	0	0	0	FR	60	0	0	100	> 1,000	24	0	0	0	0	0	100	0
8.4 ML Sensor Sys	> 1,000	24	0	0	0	0	0	100	0	24	0	0	0	100	0	0	0	FR	60	0	0	100	> 1,000	24	0	0	0	0	0	100	0
8.5 ADS	> 1,000	24	0	0	0	0	0	100	0	24	0	0	0	100	0	0	0	FR	60	0	0	100	> 1,000	24	0	0	0	0	0	100	0
9.1 MARS	> 1,000	24	0	0	0	0	0	100	0	24	0	0	0	100	0	0	0	FR	60	0	0	100	> 1,000	6	0	0.4	1,000	0	2	50	100
9.2 WAMS	> 1,000	24	0	0	1,000	0	0	100	0	24	0.1	1,000	1	100	0	1.2	2	FR	60	0	0	100	> 1,000	0	0.5	80	1,000	1,000	0	100	100
9.3 Sanctuary Base	0 - 1	0	0.5	90	1,000	1,000	2	100	100	24	0	0	0	100	0	0	0	FR	60	0	0	100	0 - 1	0	0	0	1,000	1,000	2	100	100

Figure E-6. System Scoring - *Engage* Task

163

System Name	Gulliver's	Zaibatsu	Khan	Digital	2015	Halfs
1.1 Hypersonic Attack Aircraft	24.3	21.9	24.3	20.3	22.9	24.3
1.2 Fotofighter	24.3	22.5	24.5	21.2	23.6	24.3
1.3 Container Aircraft	16.0	14.9	15.8	14.8	14.2	15.9
1.4 Lighter-than-Air Airlifter	18.2	14.2	17.8	13.7	15.7	18.2
1.5 Supersonic Airlifter	8.8	5.9	8.5	5.6	7.6	8.8
1.6 Stealth Airlifter	10.5	7.2	10.3	6.8	9.5	10.5
1.7 Global Transport Aircraft	13.1	9.1	12.7	8.9	11.5	13.1
2.1 Strike UAV	14.9	10.7	14.9	10.4	15.0	14.9
2.2 Reconnaissance UAV	26.9	30.1	26.7	28.8	25.1	26.9
2.3 Uninhabited Combat Air Vehicle	34.5	36.0	34.5	35.4	33.5	34.5
2.4 Precision Delivery System	16.4	12.0	16.0	11.5	14.6	16.4
2.5 UAV Mothership	14.6	12.9	14.3	12.8	13.4	14.6
2.6 Exfiltration Rocket	8.9	6.3	8.7	6.0	7.8	8.9
3.1 Orbital Maneuvering Vehicle	8.4	9.2	8.2	9.3	8.4	8.4
3.2 Orbital Combat Vehicle	20.0	24.0	19.9	23.5	20.1	20.0
3.3 Satellite Bodyguards	18.6	22.4	18.5	21.4	19.0	18.6
4.1 Piloted SSTO Transatmospheric Vehicle	25.5	28.5	25.4	28.0	25.5	25.5
4.2 Uninhabited Air-Launched Transatmospheric Vehicle	25.3	28.4	25.1	27.8	25.2	25.3
5.1 Adjustable Yield Munition	12.9	9.5	12.8	9.3	12.3	12.9
5.2 Advanced Air-to-Air Missile	12.9	10.4	12.7	10.3	13.2	12.9
5.3 Airborne High-Power Microwave Weapon	16.8	13.6	16.8	14.3	17.7	16.8
5.4 Standoff Hypersonic Missile	15.1	11.5	15.1	11.2	14.3	15.1
5.5 Attack Microbots	26.4	25.6	26.7	26.8	28.4	26.4
5.6 Airborne Holographic Projector	11.9	9.0	11.8	8.8	11.5	11.9
5.7 Hybrid High-Energy Laser System	19.8	18.5	20.0	18.7	21.9	19.8
6.1 Global Area Strike System	35.3	34.7	35.5	33.8	36.1	35.3
6.2 Space-Based Kinetic Energy Weapon	15.8	12.2	16.0	12.2	17.4	15.8
6.3 Space-Based High-Power Microwave Weapon	14.3	12.6	14.6	13.4	16.4	14.3
6.4 Space-Based High-Energy Laser	28.4	28.4	28.6	26.8	29.3	28.4
6.5 Solar-Powered High-Energy Laser System	28.2	28.3	28.4	26.6	29.1	28.2
6.6 Solar Energy Optical Weapon	7.8	6.6	7.8	6.7	8.3	7.8
6.7 Asteroid Mitigation System	6.0	5.3	5.9	5.3	6.1	6.0
7.1 Spoken Language Translator	11.6	11.6	11.4	11.6	10.7	11.6
7.2 Personal Digital Assistant	17.3	18.3	17.1	18.6	15.9	17.4
7.3 Virtual Interaction Center	23.2	24.2	23.0	23.9	21.3	23.2
8.1 Global Information Management System	47.7	55.5	47.8	54.2	47.0	47.7
8.2 Global Surveillance, Recon, and Targeting System	35.9	43.8	35.9	42.6	34.2	35.9
8.3 Sensor Microbots	20.9	21.5	20.9	21.9	21.8	20.9
8.4 Multiband Laser Sensor System	13.0	11.4	12.8	11.3	12.4	13.0
8.5 Asteroid Detection System	15.0	17.8	14.8	16.2	14.0	15.0
9.1 Mobile Asset Repair Station	7.7	6.4	7.4	6.4	7.4	7.7
9.2 Weather Analysis and Modification System	13.5	11.9	13.4	11.9	13.4	13.5
9.3 Sanctuary Base	47.3	44.2	47.3	42.3	44.6	47.3

Figure E-7. Final System Values By Future - AU Team Weights

System Name	Gulliver's	Zaibatsu	Khan	Digital	2015	Halfs
1.1 Hypersonic Attack Aircraft	28.9	24.4	23.9	20.3	28.9	21.1
1.2 Fotofighter	27.3	23.2	24.5	23.5	31.1	21.9
1.3 Container Aircraft	15.2	8.7	13.9	10.5	13.1	12.5
1.4 Lighter-than-Air Airlifter	19.0	13.4	16.6	7.8	15.1	12.5
1.5 Supersonic Airlifter	9.8	4.2	8.3	1.9	6.0	5.7
1.6 Stealth Airlifter	12.5	5.4	10.2	3.6	8.1	7.6
1.7 Global Transport Aircraft	14.3	5.4	13.1	2.1	9.8	9.0
2.1 Strike UAV	16.9	7.6	13.8	7.9	17.4	14.0
2.2 Reconnaissance UAV	26.8	32.8	31.2	25.2	24.5	23.5
2.3 Uninhabited Combat Air Vehicle	35.9	42.6	36.8	35.0	37.3	33.8
2.4 Precision Delivery System	17.7	7.8	17.6	3.5	10.9	11.4
2.5 UAV Mothership	13.4	9.3	14.6	6.9	12.9	11.2
2.6 Exfiltration Rocket	10.7	5.1	9.2	2.6	5.8	6.4
3.1 Orbital Maneuvering Vehicle	5.3	6.9	8.3	5.0	4.9	7.6
3.2 Orbital Combat Vehicle	14.9	18.6	20.0	18.4	16.0	17.8
3.3 Satellite Bodyguards	14.6	18.0	19.6	17.6	14.6	15.9
4.1 Piloted SSTO Transatmospheric Vehicle	19.7	24.6	28.1	21.5	20.1	22.0
4.2 Uninhabited Air-Launched Transatmospheric Vehicle	19.6	24.8	27.9	21.7	20.2	21.9
5.1 Adjustable Yield Munition	14.5	5.2	12.0	4.4	10.3	11.0
5.2 Advanced Air-to-Air Missile	10.8	6.1	15.5	5.6	12.8	11.5
5.3 Airborne High Power Microwave Weapon	17.4	11.6	17.6	11.5	18.5	18.1
5.4 Standoff Hypersonic Missile	18.3	9.0	16.2	6.0	12.9	12.6
5.5 Attack Microbots	28.4	34.6	29.9	30.0	30.0	32.9
5.6 Airborne Holographic Projector	13.2	4.6	11.7	4.0	9.3	10.3
5.7 Hybrid High-Energy Laser System	15.9	15.0	18.3	15.9	21.6	19.8
6.1 Global Area Strike System	32.3	32.9	34.4	31.3	38.6	32.6
6.2 Space-Based Kinetic Energy Weapon	15.9	9.8	14.5	10.6	19.9	15.5
6.3 Space-Based High-Power Microwave Weapon	12.7	11.0	15.2	11.2	15.9	15.8
6.4 Space-Based High-Energy Laser	28.0	31.6	30.3	27.2	31.5	26.1
6.5 Solar-Powered High-Energy Laser System	27.8	31.4	30.2	27.1	31.2	25.9
6.6 Solar Energy Optical Weapon	7.3	3.6	7.9	3.3	5.6	7.0
6.7 Asteroid Mitigation System	4.8	1.5	7.6	0.9	2.9	5.0
7.1 Spoken Language Translator	9.7	4.2	11.3	7.4	8.2	9.2
7.2 Personal Digital Assistant	14.2	6.4	15.7	16.6	13.7	14.5
7.3 Virtual Interaction Center	20.4	8.4	22.1	20.5	20.1	20.6
8.1 Global Information Management System	45.6	52.8	50.8	60.3	46.9	49.4
8.2 Global Surveillance, Recon, and Targeting System	31.2	45.9	37.5	41.8	32.5	33.2
8.3 Sensor Microbots	21.0	26.1	25.6	22.5	19.5	23.6
8.4 Multiband Laser Sensor System	11.5	8.2	17.1	6.6	9.0	10.1
8.5 Asteroid Detection System	15.3	15.9	16.3	14.0	13.6	12.9
9.1 Mobile Asset Repair Station	6.6	1.5	9.6	0.8	3.8	6.2
9.2 Weather Analysis and Modification System	12.3	6.6	13.5	6.6	10.7	11.5
9.3 Sanctuary Base	48.5	34.4	46.7	39.5	52.5	41.8

Figure E-8. Final System Values By Future - Alt Futures Weights

Appendix F

Technology Scores

This appendix gives the numerical data from the technology assessment. Table 5 shows the raw technology scores for each system. These scores were then multiplied by each system's value for a given alternate future and set of Value Model weights. The weighted technology score was computed by adding all (or top 11) of the system's weighted scores for a given technology area. Finally, the results were normalized to a 100-point scale and are presented in table 6 through table 9.

Table 5

Technology Scores by System

SYSTEM	MCTL 1.0	MCTL 2.1	MCTL 2.2.1	MCTL 2.2.5	MCTL 2.6	MCTL 4.1.1	MCTL 4.1.2	MCTL 4.1.3	MCTL 4.1.4
1.0 AIR VEHICLES (PILOTED)									
1.1 Hypersonic Attack Aircraft	20	0	0	0	0	10	0	0	0
1.2 Fotofighter	10	0	0	0	20	10	0	0	0
1.3 Container Aircraft	10	0	0	0	0	0	0	0	0
1.4 Lighter-than-Air Airlifter	40	0	0	0	0	10	0	0	0
1.5 Supersonic Airlifter	30	0	0	0	0	10	0	0	0
1.6 Stealth Airlifter	30	0	0	0	0	0	0	0	0
1.7 Global Transport Aircraft	40	0	0	0	0	0	0	0	0
2.0 AIR VEHICLES (UNINHABITED)									
2.1 Strike UAV	5	0	0	0	0	0	0	0	5
2.2 Reconnaissance UAV	10	0	0	0	0	0	0	10	10
2.3 Uninhabited Combat Air Vehicle	10	0	0	0	0	0	0	5	10
2.4 Precision Delivery System	20	0	0	0	10	0	0	0	0
2.5 UAV Mothership	5	0	0	20	0	0	0	0	0
2.6 Exfiltration Rocket	25	0	0	0	0	0	0	0	0
3.0 SPACE VEHICLES									
3.1 Orbital Maneuvering Vehicle	0	0	0	20	10	5	0	0	0
3.2 Orbital Combat Vehicle	0	0	0	20	10	5	0	0	0
3.3 Satellite Bodyguards	0	0	0	0	0	10	0	5	0
4.0 AIR AND SPACE VEHICLES									
4.1 Piloted SSTO Transatmospheric Vehicle	20	0	0	0	0	10	0	0	0
4.2 Uninhabited Air-Launched Transatmospheric Vehicle	10	0	0	10	0	10	0	0	0
5.0 AIR and GROUND-BASED WEAPONS									
5.1 Adjustable Yield Munition	0	0	0	0	0	0	0	0	0
5.2 Advanced Air-to-Air Missile	0	0	0	0	20	10	0	0	0
5.3 Airborne High-Power Microwave Weapon	10	0	0	0	0	0	0	0	0
5.4 Standoff Hypersonic Missile	10	0	0	0	0	0	0	0	0
5.5 Attack Microbots	0	0	0	10	35	5	0	0	0
5.6 Airborne Holographic Projector	0	0	0	0	0	10	0	0	70
5.7 Hybrid High-Energy Laser	0	0	0	0	0	0	0	0	0
6.0 SPACE-BASED WEAPONS									
6.1 Global Area Strike System	10	0	0	0	5	10	0	0	0
6.2 Space-Based Kinetic Energy Weapon	15	0	0	0	15	10	0	0	0
6.3 Space-Based High-Power Microwave Weapon	10	0	0	0	0	10	0	0	0
6.4 Space-Based High-Energy Laser System	0	0	0	0	0	4	0	8	8
6.5 Solar-Powered High-Energy Laser System	0	0	0	0	0	3	0	6	6
6.6 Solar Energy Optical Weapon	0	0	0	10	0	0	0	0	0
6.7 Asteroid Mitigation System	0	0	0	0	0	0	0	0	0
7.0 INDIVIDUAL INFORMATION SYSTEMS									
7.1 Spoken Language Translator	0	0	0	0	0	20	0	0	0
7.2 Personal Digital Assistant	0	0	0	0	5	10	0	10	0
7.3 Virtual Interaction Center	0	0	0	0	0	20	50	0	20
8.0 GLOBAL INFORMATION SYSTEMS									
8.1 Global Information Management System	0	0	0	0	5	5	7	4	4
8.2 Global Surveillance, Reconnaissance, and Targeting Sys	0	0	0	0	10	6	0	7	7
8.3 Sensor Microbots	0	0	0	5	35	5	0	0	0
8.4 Multiband Laser Sensor System	0	0	0	0	0	10	0	0	5
8.5 Asteroid Detection System	0	0	0	0	0	10	0	0	15
9.0 MISCELLANEOUS SYSTEMS									
9.1 Mobile Asset Repair Station	30	30	15	15	10	0	0	0	0
9.2 Weather Analysis and Modification System	0	0	0	0	5	25	0	0	0
9.3 Sanctuary Base	10	5	0	15	15	5	4	0	3

Table 5 (Continued)

SYSTEM	MCTL 4.1.6	MCTL 4.2.4	MCTL 4.2.5	MCTL 4.2.9	MCTL 4.3	MCTL 5.1	MCTL 5.3	MCTL 5.4	MCTL 5.5
1.0 AIR VEHICLES (PILOTED)									
1.1 Hypersonic Attack Aircraft	0	0	0	0	0	0	0	0	0
1.2 Fotofighter	0	0	5	5	10	0	0	0	0
1.3 Container Aircraft	0	0	0	0	0	0	0	15	0
1.4 Lighter-than-Air Airlifter	0	0	0	0	0	0	0	0	0
1.5 Supersonic Airlifter	0	0	0	0	0	0	0	0	0
1.6 Stealth Airlifter	0	0	0	0	0	0	0	0	0
1.7 Global Transport Aircraft	0	0	0	0	0	0	0	0	0
2.0 AIR VEHICLES (UNINHABITED)									
2.1 Strike UAV	0	0	5	0	0	0	0	20	15
2.2 Reconnaissance UAV	0	0	20	0	0	10	0	0	0
2.3 Uninhabited Combat Air Vehicle	0	0	10	10	0	5	0	0	5
2.4 Precision Delivery System	0	0	0	0	0	0	0	0	0
2.5 UAV Mothership	0	0	0	0	0	0	0	0	0
2.6 Exfiltration Rocket	0	0	0	0	0	0	0	0	0
3.0 SPACE VEHICLES									
3.1 Orbital Maneuvering Vehicle	0	0	0	20	0	0	0	0	0
3.2 Orbital Combat Vehicle	0	0	0	20	0	0	0	0	0
3.3 Satellite Bodyguards	0	0	20	0	0	0	0	0	0
4.0 AIR AND SPACE VEHICLES									
4.1 Piloted SSTO Transatmospheric Vehicle	0	0	0	0	0	0	0	0	0
4.2 Uninhabited Air-Launched Transatmospheric Vehicle	0	0	10	10	0	0	0	0	0
5.0 AIR and GROUND-BASED WEAPONS									
5.1 Adjustable Yield Munition	0	0	0	0	0	0	0	0	0
5.2 Advanced Air-to-Air Missile	0	0	20	0	0	0	0	0	0
5.3 Airborne High Power Microwave Weapon	0	0	0	0	0	0	0	0	0
5.4 Standoff Hypersonic Missile	0	0	0	0	0	10	0	0	0
5.5 Attack Microbots	0	0	5	5	0	5	0	5	0
5.6 Airborne Holographic Projector	0	0	0	0	0	0	0	0	0
5.7 Hybrid High-Energy Laser	0	0	0	0	0	0	0	0	0
6.0 SPACE-BASED WEAPONS									
6.1 Global Area Strike System	0	0	0	0	0	5	0	0	0
6.2 Space-Based Kinetic Energy Weapon	0	0	0	0	0	15	0	0	0
6.3 Space-Based High-Power Microwave Weapon	0	0	0	0	0	0	0	0	0
6.4 Space-Based High-Energy Laser System	0	10	10	0	0	0	0	0	0
6.5 Solar-Powered High-Energy Laser System	0	7	8	0	0	0	0	0	0
6.6 Solar Energy Optical Weapon	0	0	0	0	0	0	0	0	0
6.7 Asteroid Mitigation System	0	0	0	0	0	0	0	0	0
7.0 INDIVIDUAL INFORMATION SYSTEMS									
7.1 Spoken Language Translator	20	0	0	60	0	0	0	0	0
7.2 Personal Digital Assistant	5	0	0	10	0	10	25	0	5
7.3 Virtual Interaction Center	0	10	0	0	0	0	0	0	0
8.0 GLOBAL INFORMATION SYSTEMS									
8.1 Global Information Management System	0	5	15	15	0	8	4	5	8
8.2 Global Surveillance, Reconnaissance, and Targeting System	0	5	15	0	0	8	4	5	8
8.3 Sensor Microbots	0	0	5	5	0	5	0	5	0
8.4 Multiband Laser Sensor System	0	0	0	0	0	0	0	0	0
8.5 Asteroid Detection System	0	0	15	0	0	0	0	0	0
9.0 MISCELLANEOUS SYSTEMS									
9.1 Mobile Asset Repair Station	0	0	0	0	0	0	0	0	0
9.2 Weather Analysis and Modification System	0	0	10	20	10	0	0	0	0
9.3 Sanctuary Base	0	3	0	5	0	0	0	0	0

Table 5 (Continued)

SYSTEM	MCTL 6.1	MCTL 6.2	MCTL 6.4	MCTL 6.6	MCTL 6.7	MCTL 6.8	MCTL 6.9	MCTL 7.3	MCTL 9.1
1.0 AIR VEHICLES (PILOTED)									
1.1 Hypersonic Attack Aircraft	0	0	0	0	0	0	0	0	0
1.2 Fotofighter	0	5	0	0	0	5	0	0	0
1.3 Container Aircraft	0	0	0	0	0	0	0	25	0
1.4 Lighter-than-Air Airlifter	0	0	0	0	0	0	0	10	0
1.5 Supersonic Airlifter	0	0	0	0	0	0	0	0	30
1.6 Stealth Airlifter	0	0	0	0	0	0	0	0	35
1.7 Global Transport Aircraft	0	0	0	0	0	0	0	0	30
2.0 AIR VEHICLES (UNINHABITED)									
2.1 Strike UAV	0	0	0	0	0	0	0	0	0
2.2 Reconnaissance UAV	0	10	0	0	0	10	10	0	0
2.3 Uninhabited Combat Air Vehicle	0	5	10	0	0	5	5	0	0
2.4 Precision Delivery System	0	0	0	0	0	0	0	70	0
2.5 UAV Mothership	0	0	0	0	0	0	0	0	0
2.6 Exfiltration Rocket	0	0	0	0	0	0	0	15	0
3.0 SPACE VEHICLES									
3.1 Orbital Maneuvering Vehicle	0	0	0	0	0	0	0	0	0
3.2 Orbital Combat Vehicle	0	0	0	0	0	0	0	0	0
3.3 Satellite Bodyguards	0	0	0	0	0	0	0	0	0
4.0 AIR AND SPACE VEHICLES									
4.1 Piloted SSTO Transatmospheric Vehicle	0	0	0	0	0	0	0	0	0
4.2 Uninhabited Air-Launched Transatmospheric Vehicle	0	0	0	0	0	0	0	10	0
5.0 AIR and GROUND-BASED WEAPONS									
5.1 Adjustable Yield Munition	0	0	0	0	0	0	0	0	0
5.2 Advanced Air-to-Air Missile	5	10	5	5	0	10	5	10	0
5.3 Airborne High-Power Microwave Weapon	0	0	0	0	0	0	0	0	0
5.4 Standoff Hypersonic Missile	0	0	0	0	0	0	0	0	0
5.5 Attack Microbots	0	0	0	0	0	0	0	10	0
5.6 Airborne Holographic Projector	0	0	0	0	0	0	0	0	0
5.7 Hybrid High-Energy Laser	0	0	0	0	0	0	0	0	0
6.0 SPACE-BASED WEAPONS									
6.1 Global Area Strike System	0	0	0	0	0	0	0	5	0
6.2 Space-Based Kinetic Energy Weapon	0	0	0	0	0	0	0	15	0
6.3 Space-Based High-Power Microwave Weapon	0	0	0	0	0	0	0	0	0
6.4 Space-Based High-Energy Laser System	0	0	0	0	0	0	0	0	0
6.5 Solar-Powered High-Energy Laser System	0	0	0	0	0	0	0	0	0
6.6 Solar Energy Optical Weapon	0	0	0	0	0	0	0	0	0
6.7 Asteroid Mitigation System	0	0	0	0	0	0	0	25	0
7.0 INDIVIDUAL INFORMATION SYSTEMS									
7.1 Spoken Language Translator	0	0	0	0	0	0	0	0	0
7.2 Personal Digital Assistant	0	0	0	0	0	0	0	0	0
7.3 Virtual Interaction Center	0	0	0	0	0	0	0	0	0
8.0 GLOBAL INFORMATION SYSTEMS									
8.1 Global Information Management System	3	5	0	2	1	3	1	0	0
8.2 Global Surveillance, Reconnaissance, and Targeting System	2	10	0	2	2	9	0	0	0
8.3 Sensor Microbots	2	2	2	0	0	2	2	10	0
8.4 Multiband Laser Sensor System	0	0	0	0	0	0	0	0	0
8.5 Asteroid Detection System	0	10	0	0	0	0	0	0	0
9.0 MISCELLANEOUS SYSTEMS									
9.1 Mobile Asset Repair Station	0	0	0	0	0	0	0	0	0
9.2 Weather Analysis and Modification System	0	0	0	0	0	0	10	0	0
9.3 Sanctuary Base	0	0	0	0	0	0	0	0	0

Table 5 (Continued)

SYSTEM	MCTL 9.2	MCTL 9.4	MCTL 9.5.1	MCTL 9.5.2	MCTL 9.5.4	MCTL 9.8	MCTL 10.1	MCTL 10.2
1.0 AIR VEHICLES (PILOTED)								
1.1 Hypersonic Attack Aircraft	25	0	0	0	35	0	0	0
1.2 Fotofighter	0	0	0	0	5	0	10	0
1.3 Container Aircraft	0	0	0	0	50	0	0	0
1.4 Lighter-than-Air Airlifter	0	0	0	0	40	0	0	0
1.5 Supersonic Airlifter	0	0	0	0	30	0	0	0
1.6 Stealth Airlifter	0	0	0	0	35	0	0	0
1.7 Global Transport Aircraft	0	0	0	0	30	0	0	0
2.0 AIR VEHICLES (UNINHABITED)								
2.1 Strike UAV	0	0	0	0	0	20	0	0
2.2 Reconnaissance UAV	0	0	0	0	0	0	0	0
2.3 Uninhabited Combat Air Vehicle	0	0	0	0	0	5	0	0
2.4 Precision Delivery System	0	0	0	0	0	0	0	0
2.5 UAV Mothership	0	0	0	0	15	0	10	5
2.6 Exfiltration Rocket	0	0	0	0	25	0	0	0
3.0 SPACE VEHICLES								
3.1 Orbital Maneuvering Vehicle	0	0	0	20	0	0	0	0
3.2 Orbital Combat Vehicle	0	0	0	20	0	0	5	0
3.3 Satellite Bodyguards	0	5	0	0	0	0	0	10
4.0 AIR AND SPACE VEHICLES								
4.1 Piloted SSTO Transatmospheric Vehicle	20	0	0	0	25	0	0	0
4.2 Uninhabited Air-Launched Transatmospheric Vehicle	0	0	0	0	20	0	0	0
5.0 AIR and GROUND-BASED WEAPONS								
5.1 Adjustable Yield Munition	0	0	0	0	0	0	0	0
5.2 Advanced Air-to-Air Missile	0	0	0	0	0	0	0	0
5.3 Airborne High-Power Microwave Weapon	0	0	0	0	0	0	0	0
5.4 Standoff Hypersonic Missile	30	0	0	0	20	0	0	0
5.5 Attack Microbots	0	0	0	0	0	0	0	0
5.6 Airborne Holographic Projector	0	0	0	0	0	0	5	5
5.7 Hybrid High-Energy Laser	0	0	20	10	0	0	0	20
6.0 SPACE-BASED WEAPONS								
6.1 Global Area Strike System	5	0	5	0	10	0	0	15
6.2 Space-Based Kinetic Energy Weapon	0	0	0	0	0	0	0	0
6.3 Space-Based High-Power Microwave Weapon	0	0	5	0	0	0	0	0
6.4 Space-Based High-Energy Laser System	0	0	0	0	0	0	0	25
6.5 Solar-Powered High-Energy Laser System	0	0	0	0	0	0	0	15
6.6 Solar Energy Optical Weapon	0	0	50	0	0	0	0	10
6.7 Asteroid Mitigation System	0	0	0	55	0	0	0	0
7.0 INDIVIDUAL INFORMATION SYSTEMS								
7.1 Spoken Language Translator	0	0	0	0	0	0	0	0
7.2 Personal Digital Assistant	0	0	0	0	0	0	0	0
7.3 Virtual Interaction Center	0	0	0	0	0	0	0	0
8.0 GLOBAL INFORMATION SYSTEMS								
8.1 Global Information Management System	0	0	0	0	0	0	0	0
8.2 Global Surveillance, Reconnaissance, and Targeting System	0	0	0	0	0	0	0	0
8.3 Sensor Microbots	0	0	0	0	0	0	0	0
8.4 Multiband Laser Sensor System	0	0	0	0	0	0	25	10
8.5 Asteroid Detection System	0	0	20	0	0	0	0	30
9.0 MISCELLANEOUS SYSTEMS								
9.1 Mobile Asset Repair Station	0	0	0	0	0	0	0	0
9.2 Weather Analysis and Modification System	0	0	0	0	0	0	0	0
9.3 Sanctuary Base	0	0	0	0	0	0	5	0

Table 5 (Continued)

SYSTEM	MCTL 10.3	MCTL 11.1	MCTL 11.2	MCTL 11.4.2	MCTL 11.4.4	MCTL 12.1	MCTL 12.7	MCTL 13.3
1.0 AIR VEHICLES (PILOTED)								
1.1 Hypersonic Attack Aircraft	0	0	0	0	0	0	10	0
1.2 Fotofighter	5	10	0	0	0	0	0	0
1.3 Container Aircraft	0	0	0	0	0	0	0	0
1.4 Lighter-than-Air Airlifter	0	0	0	0	0	0	0	0
1.5 Supersonic Airlifter	0	0	0	0	0	0	0	0
1.6 Stealth Airlifter	0	0	0	0	0	0	0	0
1.7 Global Transport Aircraft	0	0	0	0	0	0	0	0
2.0 AIR VEHICLES (UNINHABITED)								
2.1 Strike UAV	0	0	0	0	0	30	0	0
2.2 Reconnaissance UAV	10	0	0	0	0	0	0	0
2.3 Uninhabited Combat Air Vehicle	5	0	0	0	0	0	10	0
2.4 Precision Delivery System	0	0	0	0	0	0	0	0
2.5 UAV Mothership	35	10	0	0	0	0	0	0
2.6 Exfiltration Rocket	0	0	0	0	0	0	35	0
3.0 SPACE VEHICLES								
3.1 Orbital Maneuvering Vehicle	25	0	0	0	0	0	0	0
3.2 Orbital Combat Vehicle	20	0	0	0	0	0	0	0
3.3 Satellite Bodyguards	20	30	0	0	0	0	0	0
4.0 AIR AND SPACE VEHICLES								
4.1 Piloted SSTO Transatmospheric Vehicle	0	0	0	0	0	0	25	0
4.2 Uninhabited Air-Launched Transatmospheric Vehicle	0	0	0	0	0	0	20	0
5.0 AIR and GROUND-BASED WEAPONS								
5.1 Adjustable Yield Munition	0	0	0	0	0	0	100	0
5.2 Advanced Air-to-Air Missile	0	0	0	0	0	0	0	0
5.3 Airborne High-Power Microwave Weapon	30	0	60	0	0	0	0	0
5.4 Standoff Hypersonic Missile	0	0	0	0	0	0	30	0
5.5 Attack Microbots	15	0	0	0	0	0	5	0
5.6 Airborne Holographic Projector	10	0	0	0	0	0	0	0
5.7 Hybrid High-Energy Laser	10	40	0	0	0	0	0	0
6.0 SPACE-BASED WEAPONS								
6.1 Global Area Strike System	0	10	0	5	5	0	10	0
6.2 Space-Based Kinetic Energy Weapon	0	0	0	20	10	0	0	0
6.3 Space-Based High-Power Microwave Weapon	45	0	30	0	0	0	0	0
6.4 Space-Based High-Energy Laser System	10	25	0	0	0	0	0	0
6.5 Solar-Powered High-Energy Laser System	40	15	0	0	0	0	0	0
6.6 Solar Energy Optical Weapon	30	0	0	0	0	0	0	0
6.7 Asteroid Mitigation System	0	0	0	0	0	0	20	0
7.0 INDIVIDUAL INFORMATION SYSTEMS								
7.1 Spoken Language Translator	0	0	0	0	0	0	0	0
7.2 Personal Digital Assistant	20	0	0	0	0	0	0	0
7.3 Virtual Interaction Center	0	0	0	0	0	0	0	0
8.0 GLOBAL INFORMATION SYSTEMS								
8.1 Global Information Management System	0	0	0	0	0	0	0	0
8.2 Global Surveillance, Reconnaissance, and Targeting System	0	0	0	0	0	0	0	0
8.3 Sensor Microbots	15	0	0	0	0	0	0	0
8.4 Multiband Laser Sensor System	50	0	0	0	0	0	0	0
8.5 Asteroid Detection System	0	0	0	0	0	0	0	0
9.0 MISCELLANEOUS SYSTEMS								
9.1 Mobile Asset Repair Station	0	0	0	0	0	0	0	0
9.2 Weather Analysis and Modification System	0	0	20	0	0	0	0	0
9.3 Sanctuary Base	10	5	5	0	5	0	0	5

Table 6

Weighted Technology Scores
(All 43 Sys, AU Students Wts)

		GULLIVER'S TRAVAILS	ZAIBATSU	KING KHAN	DIGITAL CACOPHONY	2015 CROSSROADS	HALFS and HALF-NAUGHTS
1.0	Materials	7.86	7.06	7.80	7.00	7.57	7.85
2.1	Auto of Ind Process, Sys, and Fact	0.56	0.51	0.55	0.51	0.54	0.56
2.2.1	Num Controlled Machine Tools	0.14	0.12	0.13	0.12	0.14	0.14
2.2.5	Robots, Controllers, & End Effectors	2.86	2.96	2.85	2.99	2.87	2.86
2.6	Micromechanical Devices	5.75	5.82	5.78	5.91	5.91	5.76
4.1.1	High-Performance Computing	6.12	6.28	6.12	6.28	6.12	6.12
4.1.2	Dynamic Training and Simulation	2.02	2.19	2.02	2.20	1.92	2.02
4.1.3	Signal Processing	1.85	2.10	1.86	2.08	1.85	1.86
4.1.4	Image Processing	3.90	3.97	3.90	3.92	3.83	3.90
4.1.6	Speech Processing	0.38	0.40	0.38	0.41	0.36	0.38
4.2.4	Hard Real-Time Systems	1.53	1.67	1.54	1.65	1.53	1.53
4.2.5	Data Fusion	5.18	5.80	5.19	5.74	5.22	5.18
4.2.9	Artificial Intelligence	4.34	4.73	4.33	4.78	4.31	4.34
4.3	Hybrid Computing	0.45	0.43	0.46	0.42	0.45	0.45
5.1	Transmission	2.50	2.68	2.51	2.70	2.52	2.50
5.3	Comm Network Mgmt and Control	0.92	1.06	0.92	1.08	0.88	0.92
5.4	C^3I Systems	1.43	1.45	1.43	1.46	1.43	1.43
5.5	Information Security	1.38	1.52	1.39	1.52	1.37	1.38
6.1	Acoustic Systems	0.39	0.43	0.39	0.43	0.39	0.39
6.2	Optical Sensors	1.78	2.02	1.78	1.99	1.75	1.78
6.4	Electronic Combat	0.54	0.56	0.54	0.57	0.54	0.54
6.6	Magnetometers	0.28	0.31	0.28	0.31	0.28	0.28
6.7	Gravity Meters	0.14	0.18	0.14	0.18	0.14	0.14
6.8	Radar	1.44	1.61	1.44	1.60	1.42	1.44
6.9	Other Sensor	0.88	0.93	0.87	0.93	0.87	0.88
7.3	Vehicle and Flight Control	3.94	3.46	3.90	3.48	3.83	3.93
9.1	Gas Turbine Propulsion Systems	1.23	0.87	1.20	0.85	1.11	1.23
9.2	Ramjet, Scramjet, and CC Engines	2.09	2.02	2.10	1.99	2.07	2.09
9.4	Rockets	0.11	0.14	0.11	0.14	0.12	0.11
9.5.1	Spacecraft Structures	1.60	1.60	1.61	1.60	1.71	1.60
9.5.2	Nonchem, High-Isp Propulsion	1.31	1.41	1.31	1.43	1.38	1.31
9.5.4	Aircraft High-Performance Structures	6.91	6.31	6.86	6.24	6.54	6.91
9.8	Other Propulsion	0.56	0.49	0.57	0.49	0.57	0.56
10.1	Lasers	1.33	1.27	1.33	1.26	1.30	1.33
10.2	Optics	3.64	3.80	3.66	3.71	3.78	3.64
10.3	Power Systems	8.81	8.84	8.84	8.95	9.11	8.81
11.1	High-Energy Laser Systems	4.15	4.29	4.18	4.23	4.37	4.15
11.2	High-Power Radio Frequency Sys	2.33	2.04	2.35	2.16	2.50	2.33
11.4.2	Kinetic Energy Projectiles	0.59	0.52	0.60	0.52	0.65	0.59
11.4.4	Kinetic Energy Platform Mgmt	0.68	0.64	0.69	0.64	0.71	0.68
12.1	Warheads, Annunition, and Payloads	0.54	0.40	0.54	0.39	0.55	0.54
12.7	Mil Explosives (Energetic Material)	5.27	4.89	5.26	4.89	5.22	5.27
13.3	CBW Defensive Systems	0.28	0.27	0.28	0.27	0.27	0.28

Table 7

Weighted Technology Scores
(Top 11 Sys, AU Students Wts)

		GULLIVER'S TRAVAILS	ZAIBATSU	KING KHAN	DIGITAL CACO-PHONY	2015 CROSS-ROADS	HALFS and HALF-NAUGHTS
1.0	Materials	6.10	6.01	6.08	6.01	6.02	6.10
2.1	Auto of Ind Process, Sys, and Fact	0.65	0.58	0.65	0.57	0.62	0.65
2.2.1	Num Controlled Machine Tools	0.00	0.00	0.00	0.00	0.00	0.00
2.2.5	Robots, Controllers, & End Effectors	3.39	3.14	3.39	3.16	3.37	3.39
2.6	Micromechanical Devices	6.66	6.38	6.69	6.54	6.76	6.66
4.1.1	High-Performance Computing	5.21	5.23	5.21	5.24	5.24	5.21
4.1.2	Dynamic Training and Simulation	1.45	1.47	1.45	1.47	1.42	1.45
4.1.3	Signal Processing	3.54	3.67	3.54	3.63	3.51	3.54
4.1.4	Image Processing	4.41	4.48	4.41	4.44	4.35	4.41
4.1.6	Speech Processing	0.00	0.00	0.00	0.00	0.00	0.00
4.2.4	Hard Real-Time Systems	2.88	2.90	2.89	2.85	2.90	2.88
4.2.5	Data Fusion	8.39	8.80	8.38	8.78	8.31	8.39
4.2.9	Artificial Intelligence	4.65	4.76	4.65	4.80	4.63	4.65
4.3	Hybrid Computing	0.00	0.00	0.00	0.00	0.00	0.00
5.1	Transmission	3.93	4.11	3.92	4.13	3.88	3.93
5.3	Comm Network Mgmt and Control	0.93	1.04	0.93	1.04	0.91	0.93
5.4	C^3I Systems	1.52	1.63	1.53	1.66	1.53	1.52
5.5	Information Security	2.33	2.54	2.33	2.55	2.28	2.33
6.1	Acoustic Systems	0.59	0.66	0.59	0.66	0.58	0.59
6.2	Optical Sensors	2.87	3.12	2.87	3.11	2.78	2.87
6.4	Electronic Combat	0.95	0.94	0.95	0.95	0.94	0.95
6.6	Magnetometers	0.46	0.52	0.46	0.52	0.45	0.46
6.7	Gravity Meters	0.33	0.37	0.33	0.37	0.32	0.33
6.8	Radar	2.51	2.72	2.50	2.71	2.42	2.51
6.9	Other Sensor	1.35	1.40	1.35	1.39	1.30	1.35
7.3	Vehicle and Flight Control	1.92	1.86	1.92	1.92	2.00	1.92
9.1	Gas Turbine Propulsion Systems	0.00	0.00	0.00	0.00	0.00	0.00
9.2	Ramjet, Scramjet, and CC Engines	1.90	1.94	1.89	1.95	1.93	1.90
9.4	Rockets	0.00	0.00	0.00	0.00	0.00	0.00
9.5.1	Spacecraft Structures	0.49	0.45	0.49	0.45	0.50	0.49
9.5.2	Nonchem, High-Isp Propulsion	0.00	0.00	0.00	0.00	0.00	0.00
9.5.4	Aircraft High-Performance Structures	4.14	4.24	4.12	4.27	4.20	4.14
9.8	Other Propulsion	0.48	0.47	0.48	0.47	0.47	0.48
10.1	Lasers	0.65	0.58	0.65	0.57	0.62	0.65
10.2	Optics	4.60	4.32	4.62	4.22	4.78	4.60
10.3	Power Systems	7.53	7.10	7.56	7.03	7.67	7.53
11.1	High-Energy Laser Systems	4.77	4.44	4.79	4.34	4.90	4.77
11.2	High-Power Radio Frequency Sys	0.65	0.58	0.65	0.57	0.62	0.65
11.4.2	Kinetic Energy Projectiles	0.49	0.45	0.49	0.45	0.50	0.49
11.4.4	Kinetic Energy Platform Mgmt	1.14	1.03	1.14	1.02	1.13	1.14
12.1	Warheads, Annunition, and Payloads	0.00	0.00	0.00	0.00	0.00	0.00
12.7	Mil Explosives (Energetic Material)	5.46	5.52	5.44	5.58	5.53	5.46
13.3	CBW Defensive Systems	0.65	0.58	0.65	0.57	0.62	0.65

Table 8

Weighted Technology Scores
(All 43 Sys, Alt Fut Team Wts)

		GULLIVER'S TRAVAILS	ZAIBATSU	KING KHAN	DIGITAL CACO-PHONY	2015 CROSS-ROADS	HALFS and HALF-NAUGHTS
1.0	Materials	8.34	6.97	7.69	5.97	7.58	7.17
2.1	Auto of Ind Process, Sys, and Fact	0.55	0.31	0.60	0.34	0.48	0.52
2.2.1	Num Controlled Machine Tools	0.12	0.03	0.17	0.02	0.07	0.12
2.2.5	Robots, Controllers, & End Effectors	2.68	2.85	2.88	2.84	2.76	2.91
2.6	Micromechanical Devices	5.88	6.67	5.97	6.72	6.25	6.38
4.1.1	High-Performance Computing	5.90	5.78	6.06	6.33	6.07	6.07
4.1.2	Dynamic Training and Simulation	1.90	1.32	1.90	2.43	1.96	2.05
4.1.3	Signal Processing	1.80	2.38	1.90	2.41	1.90	1.89
4.1.4	Image Processing	4.05	3.76	3.89	3.95	3.83	3.90
4.1.6	Speech Processing	0.33	0.17	0.35	0.35	0.29	0.34
4.2.4	Hard Real-Time Systems	1.50	1.74	1.52	1.96	1.64	1.58
4.2.5	Data Fusion	5.05	6.59	5.42	6.43	5.29	5.38
4.2.9	Artificial Intelligence	4.02	4.30	4.29	4.93	4.07	4.43
4.3	Hybrid Computing	0.49	0.43	0.44	0.46	0.53	0.44
5.1	Transmission	2.52	2.99	2.53	3.10	2.63	2.67
5.3	Comm Network Mgmt and Control	0.82	0.79	0.86	1.25	0.84	0.92
5.4	C^3I Systems	1.48	1.54	1.39	1.65	1.51	1.54
5.5	Information Security	1.39	1.64	1.36	1.81	1.46	1.48
6.1	Acoustic Systems	0.37	0.47	0.41	0.51	0.39	0.42
6.2	Optical Sensors	1.77	2.36	1.87	2.28	1.84	1.84
6.4	Electronic Combat	0.56	0.73	0.57	0.64	0.60	0.59
6.6	Magnetometers	0.26	0.33	0.29	0.35	0.28	0.30
6.7	Gravity Meters	0.13	0.21	0.15	0.22	0.14	0.15
6.8	Radar	1.43	1.91	1.52	1.82	1.51	1.49
6.9	Other Sensor	0.88	1.06	0.94	0.95	0.87	0.89
7.3	Vehicle and Flight Control	4.08	3.19	3.99	2.67	3.45	3.65
9.1	Gas Turbine Propulsion Systems	1.44	0.68	1.15	0.37	0.96	0.94
9.2	Ramjet, Scramjet, and CC Engines	2.27	2.19	2.10	1.93	2.16	2.00
9.4	Rockets	0.09	0.13	0.11	0.13	0.09	0.11
9.5.1	Spacecraft Structures	1.51	1.45	1.54	1.48	1.59	1.65
9.5.2	Nonchem, High-Isp Propulsion	1.03	1.06	1.35	1.02	1.01	1.30
9.5.4	Aircraft High-Performance Structures	7.28	6.14	6.58	5.27	6.44	6.08
9.8	Other Propulsion	0.64	0.52	0.53	0.50	0.68	0.60
10.1	Lasers	1.34	1.17	1.40	1.18	1.34	1.24
10.2	Optics	3.53	4.13	3.65	3.97	3.90	3.65
10.3	Power Systems	8.51	9.13	9.08	9.05	8.98	9.15
11.1	High-Energy Laser Systems	3.92	4.60	4.04	4.64	4.63	4.22
11.2	High-Power Radio Frequency Sys	2.37	1.90	2.33	2.05	2.62	2.66
11.4.2	Kinetic Energy Projectiles	0.59	0.51	0.53	0.56	0.75	0.63
11.4.4	Kinetic Energy Platform Mgmt	0.70	0.62	0.64	0.70	0.83	0.70
12.1	Warheads, Annunition, and Payloads	0.63	0.33	0.48	0.36	0.66	0.56
12.7	Mil Explosives (Energetic Material)	5.54	4.68	5.23	4.11	4.80	5.09
13.3	CBW Defensive Systems	0.30	0.25	0.27	0.30	0.33	0.28

Table 9

Weighted Technology Scores
(Top 11 Sys, Alt Fut Team Wts)

		GULLIVER'S TRAVAILS	ZAIBATSU	KING KHAN	DIGITAL CACO-PHONY	2015 CROSS-ROADS	HALFS and HALF-NAUGHTS
1.0	Materials	5.89	5.58	6.08	5.43	5.84	5.76
2.1	Auto of Ind Process, Sys, and Fact	0.71	0.44	0.61	0.55	0.72	0.61
2.2.1	Num Controlled Machine Tools	0.00	0.00	0.00	0.00	0.00	0.00
2.2.5	Robots, Controllers, & End Effectors	3.51	2.86	3.33	3.08	3.53	3.42
2.6	Micromechanical Devices	7.05	6.73	6.64	6.98	7.09	7.35
4.1.1	High-Performance Computing	4.98	4.96	5.15	5.09	5.07	5.15
4.1.2	Dynamic Training and Simulation	1.49	1.31	1.41	1.61	1.47	1.50
4.1.3	Signal Processing	3.60	3.90	3.61	3.72	3.52	3.49
4.1.4	Image Processing	4.55	4.71	4.45	4.53	4.46	4.35
4.1.6	Speech Processing	0.00	0.00	0.00	0.00	0.00	0.00
4.2.4	Hard Real-Time Systems	2.92	2.92	2.86	3.02	2.98	2.86
4.2.5	Data Fusion	8.40	9.14	8.57	8.99	8.13	8.45
4.2.9	Artificial Intelligence	4.72	4.66	4.67	5.04	4.63	4.87
4.3	Hybrid Computing	0.00	0.00	0.00	0.00	0.00	0.00
5.1	Transmission	3.97	4.29	3.97	4.30	3.86	4.06
5.3	Comm Network Mgmt and Control	0.89	1.02	0.92	1.13	0.87	0.96
5.4	C^3I Systems	1.53	1.72	1.54	1.83	1.50	1.68
5.5	Information Security	2.31	2.58	2.32	2.75	2.25	2.42
6.1	Acoustic Systems	0.58	0.64	0.59	0.73	0.56	0.63
6.2	Optical Sensors	2.87	3.25	2.93	3.18	2.71	2.87
6.4	Electronic Combat	1.04	1.10	0.96	0.97	1.02	0.99
6.6	Magnetometers	0.45	0.51	0.46	0.57	0.43	0.48
6.7	Gravity Meters	0.31	0.37	0.33	0.40	0.31	0.34
6.8	Radar	2.52	2.86	2.57	2.73	2.37	2.48
6.9	Other Sensor	1.43	1.53	1.42	1.35	1.31	1.32
7.3	Vehicle and Flight Control	1.87	1.95	1.95	1.87	1.90	2.07
9.1	Gas Turbine Propulsion Systems	0.00	0.00	0.00	0.00	0.00	0.00
9.2	Ramjet, Scramjet, and CC Engines	1.62	1.69	1.91	1.63	1.63	1.76
9.4	Rockets	0.00	0.00	0.00	0.00	0.00	0.00
9.5.1	Spacecraft Structures	0.47	0.42	0.45	0.43	0.53	0.48
9.5.2	Nonchem, High-Isp Propulsion	0.00	0.00	0.00	0.00	0.00	0.00
9.5.4	Aircraft High-Performance Structures	3.51	3.71	4.18	3.56	3.54	3.83
9.8	Other Propulsion	0.52	0.55	0.48	0.49	0.51	0.49
10.1	Lasers	0.71	0.44	0.61	0.55	0.72	0.61
10.2	Optics	4.66	4.52	4.50	4.32	5.02	4.46
10.3	Power Systems	8.00	7.66	7.61	7.29	8.13	7.61
11.1	High-Energy Laser Systems	4.89	4.54	4.66	4.43	5.21	4.59
11.2	High-Power Radio Frequency Sys	0.71	0.44	0.61	0.55	0.72	0.61
11.4.2	Kinetic Energy Projectiles	0.47	0.42	0.45	0.43	0.53	0.48
11.4.4	Kinetic Energy Platform Mgmt	1.18	0.87	1.06	0.98	1.25	1.08
12.1	Warheads, Annunition, and Payloads	0.00	0.00	0.00	0.00	0.00	0.00
12.7	Mil Explosives (Energetic Material)	4.97	5.25	5.53	4.95	4.97	5.29
13.3	CBW Defensive Systems	0.71	0.44	0.61	0.55	0.72	0.61

Bibliography

Air Force Doctrine Document: Air Force Basic Doctrine, First Draft. Washington, D. C.: Department of the Air Force, 15 August 1995.

Brassard, Michael, and Diane Ritter. *The Memory Jogger II: A Pocket Guide of Tools for Continuous Improvement and Effective Planning*. Methuen, Mass.: Growth Opportunity Alliance of Lawerance/QPC, 1994.

Clemen, R.T. *Making Hard Decisions: An Introduction to Decision Analysis*. Boston, Mass.: PWS-Kent, 1991.

Cornerstones of Information Warfare. Washington, D. C.: Department of the Air Force, 1995.

Cotton, Bob. *The Cyberspace Lexicon: An Illustrated Dictionary of Terms From Multimedia to Virtual Reality*. London: Phaedon, 1994.

Defense Planning Guidance. Washington, D. C.: Office of the Secretary of Defense, 1995.

Global Presence. Washington, D. C.: Department of the Air Force, 1995.

Global Reach, Global Power: A White Paper. Washington, D. C.: Department of the Air Force, June 1990.

Joint Pub 1-02, *DOD Dictionary of Military and Associated Terms*. Washington, D. C.: Joint Chiefs of Staff, 23 March 1994.

Joint Vision 2010, *America's Military: Shaping the Future*. Washington, D. C.: Joint Chiefs of Staff, January 1996.

Keeney, Ralph L. *Value-Focused Thinking: A Path to Creative Decisionmaking*. Cambridge, Mass.: Harvard University Press, 1992.

Kent, Glenn A. and William E. Simons. *A Framework for Enhancing Operational Capabilities*. R-4034-AF, Santa Monica, Calif.: RAND Corporation, 1991.

National Military Strategy of the United States of America. Washington, D. C.: US Government Printing Office, 1995.

A National Security Strategy of Engagement and Enlargement. Washington, D. C.: The White House, February 1995.

The Militarily Critical Technologies List. Washington, D. C.: Office of the Undersecretary of Defense for Acquisition, October 1992.

Smith, Gary R. *Logical Decisions for Windows©*. Golden, Colo.: 1995.

SPACECAST 2020 Operational Analysis. Maxwell AFB, Ala.: Air University, 1994.

USAF Scientific Advisory Board. *New World Vistas: Air and Space Power for the 21st Century,* Summary Volume. Washington, D.C.: USAF Scientific Advisory Board, 15 December 1995.

Webster's II New Riverside University Dictionary. Boston, Ma.ss: Riverside Publishing Co., 1988.

Winston, Wayne L. *Operations Research: Applications and Algorithms, 3d ed*. Belmont, Calif.: Duxbury Press, 1993.

DISTRIBUTION A:

Approved for public release; distribution is unlimited.

Air Command and Staff College
Maxwell AFB, Al 36112